THE ANTI-CATASTROPHE LEAGUE

TOM OUGH

THE ANTI-CATASTROPHE LEAGUE

THE PIONEERS AND VISIONARIES ON A QUEST TO SAVE THE WORLD

MUDLARK

Mudlark
HarperCollins*Publishers*
1 London Bridge Street
London SE1 9GF

www.harpercollins.co.uk

HarperCollins*Publishers*
Macken House, 39/40 Mayor Street Upper
Dublin 1, D01 C9W8, Ireland

First published by Mudlark 2025

1 3 5 7 9 10 8 6 4 2

A catalogue record of this book is available from the British Library

HB ISBN 978-0-00-872234-0
PB ISBN 978-0-00-872235-7

Printed and bound in the UK using 100% renewable electricity at CPI Group (UK) Ltd

'Viscous agglutinations of bubbling cells – rubbery fifteen-foot spheroids infinitely plastic and ductile – slaves of suggestion, builders of cities – more and more sullen, more and more intelligent, more and more amphibious, more and more imitative! Great God!'

H. P. Lovecraft, *At the Mountains of Madness*

'It is sometimes worthwhile to take science fiction seriously'

I. J. Goode

CONTENTS

INTRODUCTION
DEFUSING DOOMSDAY

When the Soviet Union disintegrated in 1991, the good news for Western powers was that they had lost one nuclear-armed adversary. The bad news was that they had gained several more. Soviet leaders had farmed out their nuclear, chemical and biological weapons sites to the far reaches of the empire, and many of these far reaches were now independent countries.

Those newly independent countries were politically fragile and economically troubled. And although the existence of their nuclear, chemical and biological weapons facilities had been a secret, word was quickly getting around. There is a market for enriched uranium, and the buyers are rarely the people you'd be keenest to trust with such a dangerous resource.

Andrew Weber, a young American diplomat, was dispatched to Kazakhstan in July 1993. Not much later, his car mechanic told him that there might be some uranium for sale. Via the mechanic, Weber was introduced to Vitaly Mette, a mechanical engineer and naval veteran. Mette was now the director of a metallurgical factory in Kazakhstan's desolate north-east, and he let it be known that the factory stored reactor fuel and that he might be willing to broker its sale. But he needed to trust his buyer. Over several months, Weber befriended Mette. There were a lot of dinners and a lot of vodka. On one occasion, Weber brought Mette a bottle of Wild Turkey, a rich American bourbon. Mette invited Weber to join a hunting trip in the Altai Mountains.

The trip involved a naked sauna session; par for the course in Russia. Weber shot a moose. A Kazakhstani intelligence officer, waking up in his Jeep to spot an elk walking by, shot the animal with a nine-millimetre pistol. There was much more drinking. As the trip continued, Weber slowly gained Mette's trust.

Later, at the embassy, the mechanic gave Weber another tip-off. On the snowy outskirts of Almaty, then the country's capital, the American was introduced to a former KGB border guard, Colonel Korbator. As the two men walked through the snow, Colonel Korbator discreetly passed Weber a very small piece of paper, folded in half.[1]

It was a message from Mette. Continuing to walk, Weber opened the piece of paper, read what was on it and gulped. Then he put it back in his pocket. The message was '600 kg, U-235, 90%'. The factory held not just reactor fuel, but the kind of uranium that is used in nuclear weapons. Six hundred kilograms of it – enough for about 24 nuclear bombs.

Iran was trying to get hold of the cache. North Korea and al-Qaeda were also pursuing materials of this kind. Weber was suddenly responsible for ensuring that the US's adversaries did not gain two dozen nuclear bombs. With the assistance of a chemical-nuclear engineer, Weber quietly visited the factory with Mette. There, behind a door locked by nothing but a decrepit padlock, were the uranium rods – wrapped in foil. 'This uranium metal would require nothing,' he recalled thinking, 'just being banged into the right shape and more of it – to make a bomb. It didn't need any processing. This is 90- or 91-per-cent enriched uranium-235, in pure metal form.'[2]

On receiving an urgent cable from Weber, the Washington diplomatic establishment secured the cooperation of the Kazakh government, too. 'We were able to work together,' said Weber of Mette, 'and then create what became a secret government-to-government channel. We organised a massive project to send an American team to Kazakhstan to package the materials, and then safely fly them back to the United States.'

It was a good job that Weber acted swiftly. The Americans later discovered that it was only a paperwork glitch that had prevented a shipment of beryllium, which is used in atom bombs, from being sent to Tehran.

Weber told me his story in November 2022. We spoke as part of my work for a philanthropic advice organisation that sought to find, on behalf of wealthy individuals and foundations, the best opportunities via which to address the world's gravest perils. By now, Weber was a senior fellow at the Council on Strategic Risks, a think tank that analyses and addresses many of those severe problems: climate change, nuclear weapons, and so on. He had taken this job after serving as President Obama's assistant secretary of defence for nuclear, chemical and biological defence programs.

Our conversation also touched on another Kazakhstani weapons repository – perhaps an even more harrowing one. In June 1995, Weber was one of five experts who made a secret visit to what was believed to be the world's largest biological weapons factory.

'I had been briefed on this facility,' Weber told me. 'I had seen satellite photos of the facility. In fact, the person who built the facility had defected to the United States in 1992, so we had pretty detailed information. But to walk into these buildings and to see the destructive potential, the scale ... I remember thinking at the time that Ronald Reagan was right. It had been an evil empire.'

The factory was designed to produce hundreds of tons per year of anthrax, a bacterium that causes high fever, internal bleeding and death. The world had generally maintained a taboo against biological weapons, but the Soviets had a facility that made even Weber shudder. He saw the chambers where infected animals had been imprisoned, and the vast fermenters where the bacteria would have been incubated. These horrors were being prepared for the eventuality of war with the West.

'That day and that experience was seared into me,' Weber told me. 'And while much of my career involved chemical weapons and nuclear weapons, for me, biological weapons were always the highest priority – countering biological weapons.'

Despite the work of Weber, his colleagues in the diplomatic corps and the many people who cooperated with them, the challenge of biological weapons has not gone away. In fact, as I shall explain, it has become more severe. All this is taking place before we have solved climate change and other enduring problems.

That this is all happening at once, and in the same era as political polarisation, environmental degradation and the moral abomination of factory farming, is no coincidence. Each individual crisis is downstream of what the philosopher Daniel Schmactenberger calls the 'metacrisis'. Social division has increased, because that is the natural outcome of social media companies seeking emotional engagement from their users. The planet is degraded, because that is the natural outcome of an economic system that does not price in diffuse damage. Factory-farmed animals are treated appallingly for the same reason, and the dangers of AI and synthetic biology, driven by economic incentives that reward innovators specifically and disperse the risks broadly, increase week-on-week.

None of this amounts remotely to a call to abolish capitalism and restart the Bolshevik uprising that eventually gave us the Kazakhstani caches of nuclear and biological weapons. Rather, it is to say that the things that are startlingly good about our world – technological progress, interconnectedness, and economic incentives that reward innovation – are tightly bound to the things that make it a dangerous place.

Defusing that danger, then – defusing doomsday – is a task that requires ingenuity and dexterity. Of the post-Soviet weapons round-up, Weber told me: 'In a way, we had it easy, just dealing with a known quantity, whereas the advances in science and technology are making it much, much more difficult to track and prevent the development and use of weapons of mass destruction. So we need a whole new generation of people to carry on this work.'

Over the past three years, as both a reporter and a researcher, I have encountered hundreds of the people undertaking that work. That period also included the line of work in which I met Weber. In those three years, I've met eccentric philosophical visionaries, ultra-serious diplomats, revelation-guided career-changers and brilliant technological innovators. In the field of catastrophic risk and existential risk,* there is often gloom about humanity's future. 'What's your p(doom)?' goes the common question. What that question means is: 'What probability do you put on humanity being wiped out in the next century or so?'† The natural tendency of the metacrisis is to drive our p(doom) upward, but the protagonists of this book are trying to push that number back down.

Although they would not think of themselves as such, these people are, together, an anti-catastrophe league: a diffuse team on whose shoulders rests the future of our species. Some members of this team are trying to protect us from our most ancient foes, those that are built into our Solar System and our planet. Those are the threats that our species identified first, and those are the threats that I will discuss first: asteroids, and how to avert them; volcanoes, and how to mollify them; stars, and how to withstand their rages.

In one of those endeavours in particular, we have achieved historic success. But our species' record is also one of failure: failure to keep those dangers from multiplying. As we reached for technological ascendance, we began to lose our balance.

* Those being the academic terms for things that might kill lots of us and things that might kill all of us.

† When people use the 'p(doom)' device, they're using mathematical notation. Your p(rain), for instance, might be the probability – hence the letter P – you'd assign to it currently raining. If your p(rain) = 0.01, you might well be in a desert. If your p(rain) =100, you're already wet.

Thus, in the 20th and 21st centuries, arrived the dangers of our own creation: first climate change and nuclear war, then man-made pandemics and powerful AI. I will tell the story of the singular research institute that was first to draw public attention to the most modern of those threats – before being itself destroyed. I will introduce a dark wizard of the forbidden technology that could reverse climate change. I will also give a glimpse of the plans being made for one of the worst scenarios other than annihilation: the near-total destruction of civilisation.

This is the story of the pragmatic, curious and inventive individuals who are trying to avert that destruction. Theirs is a line of work that is not always obliging; indeed, it has cost some members of this anti-catastrophe league their reputations and livelihoods.

In a broader sense, this book tells the story of a species that, already imperilled by the hazards of the natural world, has elected to hold a loaded gun to its head. Catching sight of that loaded gun in the mirror, we are now trying to dismantle it.

CHAPTER 1

THE DOUBLE ASTEROID REDIRECTION TEST

'Who knows whether, when a comet shall approach this globe to destroy it, as it often has been and will be destroyed, men will not tear rocks from their foundations by means of steam, and hurl mountains, as the giants are said to have done, against the flaming mass? – and then we shall have traditions of Titans again, and of wars with Heaven ...'

Lord Byron

There was a time when *bien pensants* scoffed at the idea that rocks could fall from the sky. Country bumpkins would occasionally report sightings of what we now know as meteorites, but men of learning dismissed these claims as fantasy. There's a rock in your field? Yes, rocks are often found in fields. The rock fell from the sky? Sounds like an old wives' tale.

Such was the intellectual consensus at the dawn of the Enlightenment. Where Romans and other pre-modern peoples had chronicled the violent arrival of stones, fragments of iron and much more, Enlightenment scientists were more sceptical.[1] It was an era in which astronomers had significantly improved humanity's understanding of our Solar System, with William Herschel, a German-British astronomer, becoming an overnight celebrity after making the first discovery of a planet (Uranus) since antiquity, and Isaac Newton developing a theory of gravity that explained planetary orbits. For all this progress, though,

European scientists generally thought of celestial bodies as processing in regal isolation rather than brutishly smashing into one another.

From this belief arose some rather over-confident comments.[2] 'It would be unforgivable to believe in such fairy tales,' said the Austrian mineralogist Andreas Stütz.[3] 'The fall of stones from the heavens is physically impossible,' stated a report that was presented, in 1772, to the Paris Academy of Sciences and was signed by the brilliant chemist Antoine Lavoisier. Ignaz von Born, the curator of the Imperial Collection in Vienna, was so afraid of ridicule that he had the museum's meteorites thrown out. When stones with black crusts were brought to them by peasants, therefore, scientists would conclude that the stones had been struck by lightning.[4]

By the beginning of the 19th century, however, opinion was beginning to shift. Ernst Chladni, a German physicist and musician, had proposed, in 1794, that the heavens were capable of flinging rock and iron at Earth.[5] Few were convinced, but the idea was nevertheless straying back towards the realm of acceptability. And so, in 1803, when provincial tales of a great fireball and a shower of stones reached the august desk of France's minister of the interior, he dispatched Jean-Baptiste Biot, a 29-year-old physicist, to investigate.[6]

Biot set about his assignment with gusto. Like a detective sent to a provincial backwater, he quickly assessed the characters of his witnesses. ('A very respectable dame, that has no interest to impress anybody,' he wrote of one; 'Concierge that seemed to be a very intelligent man,' he wrote of another.[7]) He assessed the manner in which the fragments were scattered and compared their composition to that of local stone. 'The foundries, the factories, the mines of the surroundings I have visited, have nothing in their products, nor in their slag that have with these substances any relation.'

Biot concludes his slyly florid report with what appears, initially, to be caginess. The physicist writes that his 'simple relation of facts' (so far, so guarded) would be judged best not by himself, but by 'the sagacity of philosophers ... and I shall consider myself happy if they find that I have succeeded in placing beyond a doubt' – and here is where the caginess vaporises – 'the most astonishing phenomenon ever observed by man'.

Asteroids, he concluded, were real, and the peasants correct. The extent to which academic views hinged on Biot's testimony remains contested, but there is no doubt that his was a significant contribution.[8] 'Within a few months we moved from doubt to certainty,' wrote one of Biot's peers.

We can claim Biot as a forerunner of the anti-catastrophe league, but he did not suggest that asteroids could wipe out humanity. As the doom historian Thomas Moynihan explains, the idea of human extinction was only just beginning to gestate. We were beginning to see ourselves, he writes, as 'a biological species within a desacralised cosmos', a view that was dependent on various developing strands of Enlightenment thought.[9] Over the decades that followed, evidence began to mount that the Solar System was not the wholly accommodating host that humanity had sometimes assumed it to be. Rather, it is a host that, when the mood takes it, flings rocks at its guests, sometimes rocks of dangerous size. As the astronomer Herbert Howe tartly remarked in 1897: 'While there are not definite data to reason from, it is believed that an encounter with the nucleus of one of the largest comets is not to be desired.'

Definite data began to arrive. On 30 June 1908, for instance, the silent coniferous forests of east Siberia were struck by a colossal explosion. Sixty-five kilometres (forty miles) away, Sergei Semenov, a local farmer, was sitting outside his house. He would later recall that 'the sky split in two and fire appeared high and wide over the forest ... I became so hot that I couldn't bear it, as if my shirt was on fire.'

Semenov was thrown several metres by the blast. His wife ran out of the house and led him back indoors, but the assault was not over. 'After that, such noise came, as if rocks were falling or cannons were firing, the Earth shook, and when I was on the ground, I pressed my head down, fearing rocks would smash it.'

Semenov's account was recorded by a Soviet investigator – Leonid Kulik, another enterprising mineralogist – in 1930. Kulik, visiting long after the explosion, found a wasteland the size of London: square kilometre after square kilometre of lifeless, burnt trees. Initially foxed by the lack of an impact crater, scientists would eventually piece together what had happened. It is now generally believed that the Tunguska Event, as it became known, was caused by the mid-air explosion of an incoming asteroid whose weight (100 million kilograms, or 220 million pounds) was roughly that of the Great Pyramid of Giza. The blast was the equivalent of 185 Hiroshima bombs. Its discovery was one of the more significant contributors to humanity's belated realisation that our home resembles not so much an island paradise as a coconut shy.

That you understand what is meant by the baleful unit of 'one Hiroshima bomb' is a consequence of another development of the 20th century: the creation, and use, of the nuclear bomb. These grim creations unleash an intensity of destruction that, by way of the most meagre imaginable scrap of a silver lining, gives us insight into other ways in which whole swathes of us might meet an immediate fiery end.

One of those insights has had particular significance. In the 1950s, Eugene Shoemaker, an American geologist, was studying the blast zone of a US government nuclear testing site in Nevada. It was here that he came across a curious specimen: a kind of rock that resembles quartz but is denser, more fibrous in appearance and lacking in colour and lustre.

This unusual mineral is called coesite. It's what becomes of quartz when it's subjected to intense heat (700°C, or 1,300°F) and even more intense pressure (two to three gigapascals; that is, 20,000–30,000 times the atmospheric pressure we experience at sea level). In 1960, Shoemaker and another geologist, Edward

CT Chao, found more coesite, but this time it was in a naturally occurring crater. This particular crater, which is in the desert of northern Arizona and is more than a kilometre wide, had been thought by some, given its lack of obvious celestial material, to be the product of a volcanic explosion of subterranean steam rather than of an asteroid. But in this geological whodunnit, the coesite found by Shoemaker and Chao was another smoking gun. Biot and his peers had shown that asteroids strike Earth; Shoemaker and Chao showed that asteroids strike Earth with colossal force. Along with his wife, Carolyn,* and other colleagues, Shoemaker spent much of the 1970s tracking the asteroids that hurtle around the Solar System, learning that these asteroids are also far more numerous than humanity had previously imagined.

Still, it was not until 1980 that scientists connected one dot (the enormous Chicxulub crater, in what is now the Yucatán peninsula of Mexico) with another (the sudden full stop that marked the end of Earth's aeons-long rule by dinosaurs). An asteroid had killed the dinosaurs and the remains of that asteroid were at Chicxulub. Fourteen years later came another stark warning, with humanity watching via telescope as fragments of Comet Shoemaker–Levy 9, named for the Shoemakers and their collaborator David Levy, collided with Jupiter. This was the first collision of two Solar System bodies ever to be observed, and it resulted in the eruption, from Jupiter's flank, of plumes of gas that were thousands of kilometres high.

Having hurtled through the scientific stratosphere, the idea of Armageddon-by-asteroid smashed itself deep into popular culture. Hollywood, that ever-reliable purveyor of public collective-consciousness epiphenomena, gave us the films *Deep*

* Bryan Walsh gives a fuller account of the contributions of Carolyn and Eugene M. Shoemaker in his book *End Times: A Brief Guide to the End of the World*. The Shoemakers, like Biot, deserve an honorary place in the anti-catastrophe league.

Impact (starring Robert Duvall, Morgan Freeman et al) and *Armageddon* (Bruce Willis, Billy Bob Thornton etc). These films were both released in 1998, the same year that NASA set up a division with the responsibility of monitoring near-earth objects. (Asteroids, of course, are rocky bodies that orbit the sun; near-earth objects, or NEOs, are those that soar relatively close, in Solar System terms, to Earth. While we're at it, comets are icy bodies that, when warmed by the Sun, leave a bright trail in their wake; meteors are rocky or metallic objects that, when burning through Earth's atmosphere, create characteristic streaks of light; and meteorites are the fragments of meteors that make it down to Earth.)

Over the course of this book, you will note a pattern. When a peril is first identified, it is often by someone outside scientific orthodoxy. Naturally, people scoff. (Recall here some of the earlier reactions to the emergence of COVID-19.) Eventually, it is generally accepted that a peril exists, but not that it poses an immediate threat. And then there is an incident of sufficient seriousness that it silences ridicule and turns bystanders into protagonists. In 2004, NASA forecast that the asteroid Apophis – tactfully named after Egyptian mythology's demon serpent of darkness – might hit Earth in 2029. Its later calculations were no less troubling. In December 2005, the unit of Hiroshima reared its hideous head in newspapers the world over. As a *Guardian* reporter put it: 'Nasa has estimated that an impact from Apophis, which has an outside chance of hitting the Earth in 2036, would release more than 100,000 times the energy released in the nuclear blast over Hiroshima.'[10] A later estimate, issued by a geophysicist at Imperial College London, put it at a million Hiroshimas, warning of violent ground-shaking, intense thermal radiation and atmospheric shockwaves.[11] (Even larger asteroids deliver death via monstrously sized tsunamis, depending on where they hit, and perhaps even via 'impact winter'.[12] This is a phenomenon whereby a collision kicks enough dust into the atmosphere to dim the planet for several years, drop global temperatures for the same period and starve plants of light – thereby killing them and plant-eating animals, too.)*

One of the first people to observe Apophis through a telescope was a young astronomer called Cristina Thomas. As a teenager, Thomas had been captivated by NASA's new generation of robotic spacecraft. *Cassini*, launched in 1997, sent home pictures of the rings of Saturn and dispatched a lander, *Huygens*, to the planet's largest moon, Titan. *Stardust*, launched in 1999, collected dust samples from the trail of a comet and sent that sample back to Earth. As a graduate student, Thomas analysed the composition of those small, rocky objects hurtling around the Solar System, attempting to deduce as much as possible from the limited data afforded by telescopes. Naturally, she couldn't resist swinging her telescope in the direction of Apophis. 'It was a very exciting time,' she would later tell me.

In 2006, two years after the discovery of Apophis, an updated NASA forecast showed that it would miss Earth. But the worry had resulted in a 'turning point', Thomas told me, 'when people really started to think seriously about planetary defence'.

Specifically, Congress had asked NASA, in 2005, to come up with a plan for how the agency might deflect an asteroid on a likely collision course with Earth.[13] NASA came back with some options, of which some of the most pleasing were 'Focused Solar' (use a large mirror to 'boil off' enough material to slightly alter the asteroid's orbital path), 'Mass Driver' (land on the asteroid, mine material, eject it at high velocity), and 'Asteroid Tug' (attach spacecraft and ... push!).

More practical, in NASA's view, was to detonate a nuclear bomb on, near or within the asteroid. The latter option, incidentally – drilling a hole and dropping a nuke down it – was the

* Because of the merciful infrequency of these events, there is not yet scientific unanimity on exactly what causes the temperature drop. Dust has been blamed, but there could also be a major contribution from sulphur, as a result of wildfires.

one executed by Willis and friends in *Armageddon*. But that approach might turn one big problem into many smaller ones – a host of fragments that could themselves hit Earth. 'Non-nuclear kinetic impactors,' the agency concluded, referring to the concept of deflecting an asteroid by flying a spacecraft into it, 'are the most mature approach.'

And so that was the approach which NASA tested with the Double Asteroid Redirection Test (DART), a mission in which the lead of the observations working group was one Cristina Thomas, now an assistant professor of astronomy and planetary science at Northern Arizona University. In the years that followed her early glimpse of Apophis, Thomas had retained her interest in NEOs, striking up loose collaborations with other scientists. Across the world, a network of observers was tracking asteroids together and sharing data. In 2013 came another serious incident: a house-sized asteroid exploding over Chelyabinsk, a Russian city near the Ural Mountains. The explosion generated a shockwave that shattered windows across an area of 200 square miles, and it happened to take place on 15 February. That same day, a UN group was meeting to finalise its recommendations for how Earth might best defend itself from those potentially hazardous objects that Congress had heard about.[14]

The machinery of politics was grinding into motion. The UN set up forums via which space agencies could collaborate on plans for asteroid deflection. In 2016, the US ordered NASA to create a Planetary Defence Coordination Office. Thanks to the diligent work of astronomers and the thespian labours of Bruce Willis and company,* the concept of asteroid deflection had become a focus of serious international endeavour. Thomas, having been working on NASA's observation of NEOs, was the leading astronomer advising the DART mission. 'At the time,' she

* This is enough for me to claim Willis, too, as a member of the anti-catastrophe league.

told me in 2024, 'I was kind of baffled. We had been doing this thing for fun for such a long time and all of a sudden it was real.'

The target was Dimorphos, a spheroid of compressed rubble that orbits a larger asteroid, Didymus. This ancient, lifeless pair orbits the Sun elliptically, swinging between the asteroid belt and, without threatening to collide with us, the orbit of Earth. Why, of all the countless asteroids in our vast Solar System, did NASA and the European Space Agency – an initial partner that subsequently dropped out of the mission – choose Dimorphos? One reason was that its relatively small size (roughly that of a football stadium) made it representative of the somewhat rare, but moderately risky, sort of asteroid that might hit Earth from time to time (as opposed to the very risky but very rare kind).[15] Another was the fact that Dimorphos orbited Didymus. A collision would not, according to the scientists' calculations, bump Dimorphos out of its parent's reach, which meant that it would not unintentionally put the moonlet on a radically different path that might bring it closer to Earth. Additionally, the orbital relationship – or, rather, the relative positioning of the two asteroids – makes it easier to monitor their movement.

On 24 November 2021, the DART spacecraft was launched from a site in California. The aim was to meet its quarry some 11 million kilometres (6.8 million miles) from Earth.[16] The journey took ten months and its targeting was based on the data compiled by Thomas and her fellow scientists.[17] DART was due to meet Dimorphos on 26 September 2022. It was a warm evening and hundreds of scientists and mission staff gathered at the Applied Physics Laboratory in Maryland. Most of them were under a canopy set up on the lawn. DART was equipped with a front-facing camera that was sending home pixellated footage, and that footage was being played in real time – or as close to real time as the huge distance allowed – on big monitors.

As DART soared forth at 21,000 kilometres per hour (13,000 miles per hour),[18] a clutch of white pixels at the centre of the monitor became more numerous, more grey. It was Dimorphos and it was getting larger and larger. 'We can see it getting closer,' recounted Thomas, 'and that it was going to hit Dimorphos right in the centre, which is what we wanted. We're making all these estimates in our heads of what we think is going to happen.'

As DART approached, Thomas and her fellow scientists could see that the surface of Dimorphos was covered with regolith: easily dislodged rock and dust of the kind that the Apollo astronauts walked on. 'This was great,' said Thomas, 'because of the momentum it adds when the craft strikes.'

Closer and closer came the asteroid until, abruptly, it filled the screen. Then, just as abruptly, the screen went orange. Everyone cheered. 'It was phenomenal,' Thomas told me.

Their work would have been for nothing, though, had they failed to deflect the asteroid. Across the parts of the world that had sight of the collision zone, dozens of telescopes were being moved into position. Their operators were part of the global network of NEO monitors and their data was now the most sought-after in astronomy. With NASA broadcast commentators still burbling away, Thomas took out her phone to check with her global collaborators. At her shoulder gathered colleagues and members of colleagues' families, all desperate to know what had happened. 'Minutes later, we started getting these videos from telescopes in Israel, in South Africa, because the impact was over the Indian Ocean and they were the only ones who could see it.'

The videos showed two dots moving across the sky – 'and then,' said Thomas, 'this poof of material comes off'. This was the ejecta – the material knocked off Dimorphos by the force of the impact, meaning that the asteroid had been nudged off its path and that the mission had been a success. 'I thought that, as we watched things happen, it was going to be a slow burn. I wasn't expecting it to be so visually captivating.'

Did a rager ensue? 'By scientific standards, potentially,' said Thomas, wryly. If nothing else, there were mission-themed cupcakes. One hopes that if the scientists had deflected an asteroid on course to wipe out civilisation, they'd have felt entitled to at least a beer.

There was no doubt that DART had hit Apophis, but what was less clear was the extent to which the spacecraft had altered the moonlet's trajectory. NASA did not, therefore, make any immediate claims in this regard. Any chump of a humanoid species can sling a spacecraft across a distance of 6.8 million miles to hit an asteroid squarely between the eyes.* But can they tangibly change that asteroid's path? Recall that we can calculate an asteroid's trajectory only through glimpsing it, a tiny white dot, through the sights of a telescope, and noting its position relative to other tiny white dots. The data trickled in, and the error bars shrank and shrank, but only a small number of people were privy to all the figures; only a small number of people knew that humanity had, for the first time, non-trivially altered the movement of a celestial body.

By now, Thomas had returned to her teaching duties at Northern Arizona University. 'I was one of the first people to know,' she recalled, 'and I'm walking around thinking, *I know something really, really exciting that I can't tell you!* It's not like it was a state-level secret, but it was still something I was supposed to keep to myself.' She laughed at the memory. There must be something absurd about exerting God-like power on the universe one week and going back to work the next.

But on 11 October 2022, a little more than a fortnight after the impact, NASA confirmed the successful alteration of Dimorphos' orbit.[19] It was the first full-scale demonstration of asteroid deflection technology.†

* Indeed, the DART mission wasn't even our first time doing this. That was in 2001, when NASA landed a spacecraft named after Eugene Shoemaker on the asteroid Eros.

† My telling of this story has focused on NASA, but it would be remiss of me not to mention that it was Johns Hopkins Applied Physics Laboratory that built and operated the DART spacecraft.

As the agency explained, it had previously defined success as altering the duration of Dimorphos' orbit of Didymos by 73 seconds or more.[20] The early data, however, showed that DART surpassed this minimum benchmark 25 times over. Our species had disrupted the ancient clockwork of the Solar System and was at last a protagonist, in orbital terms, rather than a passenger. 'This is a watershed moment for planetary defence and all of humanity,' said NASA administrator Bill Nelson.

A watershed though the DART mission was, humanity is not in danger of being imminently eradicated by an asteroid. Asteroids of a diameter of about one kilometre (0.74 miles), far larger than the 20-metre diameter of the Chelyabinsk asteroid,[21] are large enough to cause massive devastation, but they are estimated to hit the planet roughly once every 700,000 years. Asteroids of a five-kilometre (three-mile) diameter are estimated to hit us roughly once every 30 million years.[22] And the larger they get, the more unlikely their arrival. Given that the asteroid thought to have wiped out the dinosaurs is believed to have been 10–15 kilometres wide, we ought to have a few years' grace before the next one.

And we will see it coming. 'The truth of the matter,' said Thomas, 'is that we actually know the exact positions and essentially the entire population of the largest objects in the near-earth object category.' In other words, NASA knows who they are and where they live. It's the smaller ones which pose more of a risk, said Thomas; those which are more like the size of Dimorphos. These cause regional devastation, not global devastation, 'which, of course, is still bad! We still want to mitigate those risks.' And so the monitoring and classification continues, with dozens of observatories, large and small, supplying data. Together, they comprise an anti-catastrophe league of their own. To be contributing to this mission must be conducive to some level of job satisfaction; surely one might be tempted, when asked what one does for a living, to say: 'Well, I defend the planet from asteroids, actually. How about you?' Thomas, however, describes herself simply as a 'professor at the local university. Much more low-key. But my husband reminds me sometimes that my job is way cooler than I make it sound.'

Humanity's successful nudging of Dimorphos is a totemic success in its own right. It has also occurred ahead of schedule. We are years behind where we ought to be in the race to rid ourselves of fossil fuels, yet we have deflected an asteroid many millennia before we will find ourselves in genuine need of the technology. (It is worth noting that the DART mission, though executed with the collaboration of many space agencies, did not have to seek the approval of every man, woman and child on Earth. The fewer the stakeholders, the more efficient the process.)

Some scientists and commentators, including the great astronomer Carl Sagan, have pointed out that humanity could use its new power maliciously – that malevolent humans could deflect an asteroid towards Earth, or towards an enemy country, rather than away from it.[23] One reason to be relatively optimistic is such a projectile will not come from nowhere, and that our ability to monitor NEOs is likely to improve. Another is that our defensive abilities might improve too.

In that vein, you might remember the 'Focused Solar' plan that was mooted by NASA – the one in which a spacecraft-mounted mirror was pointed at an asteroid, burning off enough regolith to tweak its trajectory. In time, we might refine that technique. In 2012, five academics put forward a blueprint for 'laser bees', a swarm of laser-bearing spacecraft that would work together to drive away an asteroid.[24] The laser bees would be cheap to produce and would be numerous enough that the loss of one or two of them would not jeopardise their mission. Perhaps a defensive-minded planet could retain a few thousand of these in various positions across a Solar System. As SpaceX brings down the cost of space travel, such schemes will go from fanciful to viable to mundane.

Think back to that footage of Dimorphos and its dusty surface. That dust, and all those pockmarks that one sees on

Mars, the Moon and every solid object beyond Earth, are the result, by and large, of billions of years of collisions. Such collisions feel infrequent to a blink-of-an-eye species such as our own, but they are a fact of life in this Solar System and any other. For almost all of humanity's existence, the briefness of our lives has been our only shield. Now, though, we have a shield of our own making.

CHAPTER 2
A RAGING STAR

'... unruly sun,
Why dost thou thus,
Through windows, and through curtains call on us?'

John Donne

John Kappenman did not relish what he was about to do. He did not class himself as a confrontational man. Nor did he class himself as a man particularly enthusiastic about risking his income. He had spent decades advising the US energy sector and he knew that what he had written could make him a pariah.

But it would be even worse, he believed, to stay silent. So he stuck to the plan, finished his testimony and sent it to the Federal Energy Regulatory Committee. The testimony, he hoped, would hit like a thunderbolt – and shock a somnolent government into action.

Ninety-three million miles away, the Sun was seething. The Sun, to be clear, is always seething. About 4.6 billion years ago, a giant cloud of gas was drawn together by gravity, becoming so dense that nuclear fusion began to occur at its centre. Nuclear fusion is when two or more nuclei, in this case hydrogen nuclei, are fused into a new element – helium – while releasing colossal amounts of energy. The star's core reached 15 million degrees Celsius (27 million degrees Fahrenheit). Several billion years later, it remains at this temperature and the heat drives vast movement of plasma within the body of the Sun. This movement of plasma generates strong magnetic fields, which together constitute the 'solar dynamo'. (Do keep this roiling magnetism in mind.)

Above the plasma is the Sun's surface. It is hundreds of thousands of miles from the star's core, yet it is bright enough for Earth dwellers, themselves 149 million kilometres (93 million miles) away, to find it unbearable to look at. Earth captures a minuscule fraction of the light that radiates from the Sun, yet it is sunlight that sustains almost every activity on our planet's surface. Even fossil fuels are, one way or another, solar batteries. To remember that the sunlight illuminating a room is the product of unstoppable nuclear fusion, almost a hundred million miles away, is to be starkly reminded of the universe's titanic strangeness – a strangeness from which we are usually insulated. The energy arrives predictably and continuously, like energy from a power grid.

Or rather – it's *mostly* predictable and continuous, just as energy from a power grid is *mostly* predictable and continuous. The Sun is constantly expelling charged particles, but we are usually protected from this 'solar wind', as it is known, by Earth's magnetic field. As a result of its magnetic power, this field deflects the charged particles, buffeting them away from the equator. Most fly past Earth, although some hit the upper atmosphere around the North and South Poles. When this happens, the charged particles react with gases in our atmosphere to give off ethereal, rippling light shows: the Northern and Southern Lights. The light is greenish-yellow when charged particles hit oxygen atoms at a lower altitude, and redder when this happens at a higher altitude. Nitrogen produces reds and blues.[1] When an aurora appears to dance, this is largely because of fluctuations in Earth's magnetic field.

Every now and then, the wind becomes a gale. Inside the Sun, the flows of plasma and magnetism occasionally become tangled. When this happens, dark spots appear on the Sun. These are sunspots and they spell danger – a scowl on the face of an ancient god. Out of sunspots erupt solar flares, the largest explosive events in the Solar System. Solar flares blast electromagnetic radiation into space in enormous quantities. (The biggest solar flare ever measured was so powerful, reports NASA, that it overloaded the sensors measuring it.[2]) The radiation travels at the speed of light, but it is swiftly followed by charged particles that have emerged from the plasma. That is what's known as a coronal mass ejection: an expulsion of matter from the sun's corona, which is its outermost layer. If the coronal mass ejection is large enough, it won't bounce off Earth's magnetic field – it will penetrate it. By now, we're no longer talking solar winds, but solar storms. The scowling god has hurled a thunderbolt.

These events leave far fewer obvious traces than an asteroid does. It is therefore harder to know how strong these solar storms can be and how often we should expect storms that are worth worrying about. This state of affairs goes some way toward explaining why John Kappenman, the electrical engineering expert I introduced at the beginning of the chapter, so often finds his warnings falling on proverbial deaf ears.

But we have one key data point above all others: the Carrington Event. On the morning of 1 September 1859, an amateur British astronomer, Richard Carrington, was going about his customary observation of the forms and positions of sunspots. Carrington was a hobbyist *par excellence*. He lived in Redhill, Surrey, and had added a bulbous observatory dome, complete with telescope, to the roof of his home.[3] As usual, Carrington was projecting the Sun's image onto a plate of glass that was covered in a thin coating. He was counting sunspots, of which there was already a notably bigger mass than usual, when something startling occurred: the appearance of two patches of intensely bright and white light. These patches were so bright that Carrington's first assumption was that direct sunlight was illuminating his plate of glass. But this was no optical illusion.

In a development that should delight hobbyists the world over, Carrington had made the first recorded observation of a solar flare.* What he was watching was a huge and violent eruption of solar gases, an explosion roughly equivalent to the detonation of ten billion Hiroshimas.[4] Although Carrington couldn't have known it, the charged particles were now hurtling towards Earth.

Eighteen hours later, the bombardment began. The northern and southern lights illuminated skies that they had no business illuminating. Gold miners in Colorado's Rocky Mountains, assuming that the bright sky meant that morning had arrived, began preparing breakfast at 1am.[6] The northern lights were seen as far south as Colombia, provoking both wonderment and fear.[7] In Australia, far north of where the southern lights should appear, another group of gold miners looked out of their tent one evening and saw 'a scene of almost unspeakable beauty ... Lights of every imaginable color were issuing from the southern heavens.'[8]

One man's ethereal beauty is another man's harbinger of doom. On 5 September, the *San Francisco Herald* bore an account rather less grateful than that of the Australian miner. 'Half-past eleven,' it read. 'The appearance now is positively awful. The red glare is over houses, streets, and fields, and the most dreadful of conflagrations could not cast a deeper hue abroad.'[9]

This coronal mass ejection amounted to more than a light show. By now, the world was beginning to communicate via telegraph – a transformative invention that allowed people to send brief messages via an electric current sent along a wire. Because these wires were able to conduct an electric current,

* Credit is also due to another amateur astronomer, Richard Hodgson. Doing a similar survey of sunspots, he saw, by his account, 'a very brilliant star of light, much brighter than the sun's surface'.[5] Since there are no particular restrictions on anti-catastrophe league headcount, Hodgson and Carrington can both have a place.

they could also have an electric current induced in them. And because the coronal mass ejection had sent ripples through Earth's magnetic field, those wires were suddenly overloaded. At some telegraph stations, the induced current was strong enough to start fires. Several telegraph operators were burnt. At others, operators found that they could communicate without their machines having to be connected to batteries. Some of these machines began to spew gibberish. To see such a sight under a pulsing red sky must have been a deeply unnerving experience. As Carrington's biographer, Stuart Clark, wrote in 2007: 'Man's dominion suddenly felt less assured. The Earth had been caught in something inexplicable, the first phenomenon seen that had a direct influence on Earth HAD had nothing to do with gravity – the only force thought capable of communicating itself across space.'[10]

Carrington was one of the first to suggest that the violent activity he had seen on the Sun might have something to do with the occultish happenings on Earth. The strange phenomenon that the world had experienced would therefore become known as the Carrington Event. Nevertheless, Carrington was not the only gentleman scientist to identify this link between the Sun and Earth. Another was Edward Sabine, a swashbuckling Anglo-Irish astronomer. As a young man, Sabine helped resist the American advance on Quebec in the war of 1812.[11] After peace was declared, he embarked on Arctic expeditions that sought to find the Northwest Passage; sailed to the coasts of Africa and North America, and back to the Arctic, in order to study Earth's shape via use of a pendulum; and was first to observe that the cyclical behaviour of sunspots, whose aggression peaks every 11 years or so, matches the cyclicality of magnetic disturbances on Earth. Geomagnetic storms* do not occur with metronomic 11-year regularity; big geomagnetic storms can occur outside a solar maximum, as the peaks of

* A term for the electromagnetic havoc caused on Earth by a solar storm.

sunspot activity are known, and a solar maximum will usually pass without giving us anything like a Carrington Event. But we know, by and large, when to expect more volatile solar weather. The efflorescence of the northern and southern lights in the spring of 2024, for instance, was in keeping with the 11-year cycle and did not come as a surprise to astronomers.

That said, there is much about solar weather that scientists do not know. Most of the uncertainty relates to the Sun's internal activity. We have no apparatus that we can send into it in order to ascertain the direction of its plasma currents. Instead, we must study its surface – much as Carrington did. As I was told by Andrew Richards, formerly a severe events forecaster at the National Grid: 'The problem, as I see it, is that the state of solar science is not yet well enough developed for us to be able to predict when one of these explosions on the Sun is going to happen.'

There is additional, related, uncertainty. How often can we expect a geomagnetic storm of the Carrington Event? It's hard to know, but analysis of ancient tree rings found that storms significantly larger than the Carrington Event have hit Earth within the past 10,000 years.[12] And if one befell us – how bad would it be? It is now that we return to John Kappenman, who was preparing to warn a US government agency about the seriousness of geomagnetic storms.

That was in 2012. It was an intervention whose origins lay in solar cycle 22, which culminated in 1989.*[13] A pair of coronal mass ejections, one launching two days after the other, caused circuit failures in satellites and a power grid blackout in Quebec, Canada. For nine hours, millions of people were left without electricity. Kappenman, then a 34-year-old electrical engineer, was working for a big utility company, helping it build one of

* Solar cycle 1 is not, of course, the first solar cycle to have existed. It began in 1755, and it marks the first period for which Johann Rudolph Wolf, an enterprising Swiss astronomer and mathematician working in the 1840s, found sufficient sunspot data to identify a solar cycle.

the longest high-voltage lines in the world. He and his colleagues had noted the damage caused by the geomagnetic storm, knew about the Carrington Event, and were conscious that there was no obvious reason why there couldn't be bigger geomagnetic storms in future. As such, Kappenman recalled, they decided that they might need 'to do six months of work, figure out these problems and solve them'.

The six months of work turned into an odyssey of study and advocacy. It led Kappenman to write, in that testimony delivered to the Federal Energy Regulatory Committee, that geomagnetic disturbances could be ten times as intense as US government agencies believed, that we had become far more vulnerable to these events as our dependence on electronics increased – and that millions of lives were at risk.

As he explained in the docket, each major geomagnetic storm has inflicted 'new and unpleasant impacts' on electric power grids. 'Time and again,' Kappenman warned, geomagnetic storms have 'brushed away rosy assumptions and replaced them with less pleasant realities'.

By Kappenman's forecast, it was 100 per cent certain that Earth would, sooner or later, be blasted by a geomagnetic storm ten times larger than the one that struck Quebec in 1989. Such a storm, in his view, would have appalling consequences, most of which would relate to the electric power grid being put out of action. One key vulnerability is our reliance on transformers, which are large, cumbersome bits of equipment that connect power stations to power grids. Their role is to step up the voltage of the electricity that they are dispatching over long distances. This practice lowers the amount of power that's lost along the way, but the high voltage makes power lines extremely dangerous. (Birds can safely roost on them only because they aren't connected to the ground.) At the end of the high-voltage power line, another transformer steps the voltage back down, ensuring that it is at an appropriate level for distribution to homes and workplaces.

Normally, this system works very well. The Sun beams down on a solar farm somewhere; moments later, I might use some of that same energy, transported at the speed of light over a long distance, to boil a kettle. Add a geomagnetic storm, though, and things get more complicated. Recall the telegraph lines from the Carrington Event. They were conductive enough for the storm to induce in them much stronger currents than they were designed to convey. A modern transformer, however, bears on its back an even larger target. These enormous machines, which can weigh as much as a blue whale,[14] induce voltage through the rapid oscillation of their magnetic field. Add a powerful, intruding current and the magnetic core becomes overloaded. In severe instances, the transformer will rattle, overheat and produce the scents of burning insulation or vaporised oil.[15] 'They make absolutely horrible noises when they're doing this,' an expert told me.

In the worst of these instances, the transformer will break down. Want a new one? It might take several years and several million pounds, and it will have to be painstakingly transported via closed roads on the trailer of a special vehicle that is even more cumbersome than the contraption it is carrying. Perhaps understandably, power grid companies are disinclined to commission back-up transformers unless they really feel they ought to.

But perhaps we might one day find ourselves in dire need. Knock out a few transformers at once and a country could be crippled. Grid power enables not only heating, lighting and so on, but other key functions – the distribution of drinking water, the refrigeration of food, many methods of healthcare, even the avoidance of nuclear meltdown. Satellites, too, would be badly affected; we could lose access to GPS and other important communications systems.* In 2022, a solar storm destroyed 40 SpaceX satellites.[17] More troubling still was the incident in 1967

* As the American essayist Kathryn Schulz pointed out in 2024, 'only retroactively did it become apparent how lucky NASA was that no such storms happened during the Apollo missions'.[16]

when a solar storm jammed all three radar sites used by the US Air Force to detect incoming ballistic missiles. Interpreting the jamming of the systems as a Soviet attack, American military officials were close to launching a nuclear counter-strike. A small team of solar forecasters is said to have convinced the military officials that the culprit was not the Russians, but the Sun. This was an intervention that researchers have since credited with having 'calmed nerves and allowed aircraft engines to cool as they returned to normal alert stance'.[18] The episode minted several new members of the Anti-Catastrophe League,* but it serves as a reminder of the way that danger, be it from solar storms or other hazards, can rapidly cascade.

From all this, you might conclude that a massive solar storm could send us back to the Carrington era, or worse. In contrast to the Victorians, most of us would have no wells to drink from, no horses to ride on, no printed maps to navigate by and no fireplaces via which to warm our starving bodies via the reluctant incineration of our furniture. 'The recovery could plausibly extend into months, or perhaps years,' wrote Kappenman, warning that millions of lives were at risk. The threat, he argued, had been met not only by inertia, but by the power companies' attempts to suppress warnings. Comparing the situation to the Silent Spring scandal of the early 1960s, in which American chemical companies covered up the carcinogenic effects of the pesticide DDT, Kappenman wrote: 'We have a similar cloud of misinformation, inattention and even attempts on the part of some to stop any action, claim that the problem is overblown, uncritically accept industry assurances that no problem exists, or that the problem does not apply to them and even take steps to conceal any information that exists on the full nature of the threat. We must not fail to heed the warnings of history in this endeavor.' Governmental docket filings are often dry;

* It's high time that Colonel C. K. Anderson, Major Donald Sherry and Captain Lee Snyder received their flowers.

Kappenman's was the equivalent of being grabbed by the collar and told, with great urgency, that you are in grave danger.

Kappenman's was the direst warning, but he was not the only party to call attention to the threat. In 2008, the National Academy of Sciences, an American non-profit that brings together leading researchers, published a report whose authors envisaged disruption on a similar scale.[19] The researchers, partly informed by Kappenman, feared that a severe geomagnetic storm could cost the US between $1 trillion and $2 trillion during the first year alone.

In 2016, the body that Kappenman had addressed – the Federal Energy Regulatory Committee – adopted new regulations that required American power companies to test their systems for vulnerability to geomagnetic storms.[20] Across the Atlantic, the UK had opened a space weather forecasting centre that would work in conjunction with an American counterpart. Was this a happy ending? Is the world safe? Could Kappenman, having followed our asteroid-spotters in raising the alarm about an unfamiliar threat, sit back in comfort, his work done? Would Bruce Willis be starring in a Hollywood movie about ingeniously insulated transformers?

Alas, when I spoke to him in 2024, Kappenman told me that the rules were nowhere near stringent enough. Getting a grid started after a shutdown is called a black start and Kappenman, by now 70, believed that even natural gas plants would, in the modern era, rely on mains electricity to get going. 'We are creating a self-reinforcing scenario,' he warned.

Issuing these warnings, Kappenman said, has cost him. Again and again, he has called for tighter regulation on electric utility companies, travelling across the country to give evidence on his own dime. When he wasn't campaigning, he worked as a consultant, advising those companies on how best to protect their equipment against geomagnetic storms. As a result of his campaigning, he said, his business dried up. 'There's not a US electricity company that will hire me any longer,' he told me.

His wife, he said, has been 'wonderfully supportive'. Despite the lost income, 'we'll get by, we're just fine. And I guess I'd rather do that' – continue to call for more stringent safety standards – 'and try to head off this problem, than ignore the problem and maybe gain some short-term financial benefit from it.' Not every prophet of doom ends up getting a comet named after themselves. Most must suffer the dismissiveness of their peers; the dismissiveness is often correct.

Kappenman still believes the threat is 'existential'. By this, he means not that everyone will be immediately killed, but that modern society, so intimately dependent on 24/7 electricity, might not manage to rebuild in the years following a severe geomagnetic storm. We'd be sent back to the Dark Ages, but without a torch, and without even a candle. To believe the world to be so vulnerable must be a terrible burden, not least because, as a result, Kappenman has suffered some level of professional ostracism. I wonder whether this stuff keeps him up at night. 'I try to get a good night's sleep,' said Kappenman, but he hadn't achieved one the night before we spoke. He's a volunteer firefighter in Minnesota and the computer systems supporting the local 911 emergency hotline had gone down. As a consequence, the volunteers were called into the local fire hall overnight. This was so that members of the public, should they find themselves in need, could drive over and get immediate help. But it was far less safe a situation for the public than a working hotline would have enabled and it was far less comfortable for the volunteer firefighters. It later emerged, Kappenman told me, that a software update had crashed the system. In his view, the episode typified our over-reliance on fragile, centralised electronic systems. 'We haven't learned our lesson on this thing,' he said.

Kappenman has been called both a visionary and an alarmist.[21] 'He's very passionate about this,' said Chris Beck, who helps advise governments on the security of energy infrastructure. 'He's very strident and he gets upset when people don't take it as seriously as he does, and I think that's alienated some people. But, you know, I think his science is right.'

Others disagree. 'He over-exaggerates the threat,' said Andrew Richards, a power grid expert who was, for eight years, a severe risk analyst at the UK's National Grid. Richards has been following Kappenman's work since the 1990s. 'We have been at odds over the years on this,' Richards told me.

All parties seem to agree that Earth will, at some point, be hit by a storm that's akin to the Carrington storm, perhaps stronger. Richards tells me that he wishes that solar science were sufficiently well developed that we could predict eruptions on the Sun's surface and act accordingly. But where he differs from Kappenman – and where other members of the industry differ from Kappenman – is that he has a higher opinion of the resilience of existing technology.

Does Richards roll his eyes at solar storm doomerism? 'I think it's a good thing that people are aware that there is a risk,' he says, 'because there *is* a risk. Treated properly, I think it is a controllable risk and not completely out of step with other technological risks that we face. I think John's view of it is extreme. We will eventually have another large storm and I guess the proof of the pudding will be in the eating.' Richards tells me that he is 'as confident as he can be' that, in the UK, the effect of a severe solar storm would be little worse than a severe terrestrial storm.*

* Similarly, the non-governmental American entity that sets power grid standards expressed confidence when I asked them about their preparedness. The entity is called the North American Electric Reliability (NERC). It is not to be confused with its overseer, which is the Federal Energy Regulatory COMMISSION (FERC). FERC is the organisation to which Kappenman wrote. NERC ≠ FERC. Anyway, Mark Olsen, who is NERC's manager of reliability assessments, detailed to me some of the considerations underlying the organisation's regulations. He told me that while transformer damage was a concern, 'it has been overstated in reports that assume a GMD [geomagnetic disturbance] scenario will cause all transformers to fail across a wide area (and thus lead to an economic and social catastrophe)'. Overall, he said, scenarios of 'catastrophic transformer damage' were 'implausible'.

I asked National Grid about their preparedness strategy and was told the following: that the organisation 'does hold spare transformers for ... the event that a replacement is required – but we don't share the specifics'.

The spokesman did not explain the secrecy. One can surmise, though, that it has to do with national security. Power grids have several potential foes, of which the Sun is but one. Outages of the kind experienced by John Kappenman's fire crew can be malicious as well as accidental, for enemies of the West are not blind to our modern tendency to centralise and digitise. It is a tendency that helps make our infrastructure a more inviting target to cyber-attackers. Hackers linked to the Russian state have been blamed for an attack that, in the first weeks of the war in Ukraine, came close to knocking out a big chunk of the American power grid.[22] In these circumstances, it makes sense to wreath one's back-up systems in the fog of war.

Far worse than a hack would be an electromagnetic pulse attack (EMP). No EMP attack has yet been made by one country on another, but they are a troubling prospect – not least because they expose the same vulnerabilities as geomagnetic storms do. Detonate a nuclear weapon high above Earth's surface and you will send gamma rays smashing into air molecules. These collisions eject electrons from the air molecules, and these electrons, accelerated by Earth's magnetic field, blast outwards in the form of a powerful electromagnetic field – the electromagnetic pulse.

Imagine that you're at the wheel of a Tesla when a weapon of this kind is detonated. You'd hear the bang, but you wouldn't feel the electromagnetic pulse as it rushes through your body. By and large, it wouldn't affect you. But it would affect your car, which is full of circuitry designed to conduct electricity. Depending on your location, your car could be rendered completely inoperable. No dashboard screen, no power. As your car trundles to a stop, you point it towards the hard shoulder. Everyone else is doing the same. You pull your phone out of your pocket. The phone is dead too. Your confusion turns to horror as you realise that every piece of electronics within a hundred miles might now be inoperable. It's been forecast that dams will burst, planes will fall out of the sky and trains will smash into each other, just as they might in the event of a colossal solar storm.[23] An extended period without grid power would also raise the spectre of nuclear meltdown. If you know anything about transformers, you pray your national power grid has a few back-ups.

The field of catastrophic risks contains some nasty surprises, but it also contains a fortunate principle. Just as apparently discrete hazards can cause cascading vulnerability, so can solutions make us safer in more than one regard. Because EMPs are, in physical terms, similar to geomagnetic storms, our insulation against one can help insulate us against the other. In other words, it's a buy-one-get-one-free insurance deal. An EMP, detonated above the UK today, would be a catastrophe. But Chris Beck, the energy infrastructure advisor who was supportive of John Kappenman, was surprisingly optimistic about our ability to protect ourselves from EMPs. Beck, who is chief scientist and vice-president for policy at the Electric Infrastructure Security Council, invited me to recollect how analogue radios used to pick up signals that were being sent to one's mobile phone. (I immediately recalled the sonic throb via which a radio used to herald the arrival of a text message.) To prevent electronics from zapping each other, Beck explained, electrical engineers came up with innovative ways of shielding devices from unexpected charges. 'It's just a question of simple electromagnetic shielding, filtering and proper grounding,' said Beck, 'and EMP is just a little bit higher intensity. But it's the same principle.' Military equipment is already protected in this way, he explained. Giving every other bit of electronic equipment the same treatment of 'hardening', as it's known, would be expensive and inconvenient, but nevertheless feasible. EMPs, like solar storms, need not spell doom. 'If we just do the design codes right and just start building things that way, then it won't be a problem for ever.'

For the time being, Beck was banking on himself and his family being able to rely, in a pinch, on their domestic solar array and their supplies of food and water. 'We could probably stick it out for a month if we had a blackout,' he said. Kappenman hasn't bothered doing much prepping; he thinks the problems caused by a severe geomagnetic storm would be so enormous that prepping wouldn't be much help. 'There are certain things that can't be done, especially if you're dependent on life-saving medications. You can't stockpile a decade of those.'

Kappenman might be heartened, if only briefly, by the progress of Ahmad Kirmani, a materials scientist at Rochester Institute of Technology. Kirmani has been developing materials called perovskites, which can recover after being damaged by radiation.[24] Maddeningly, perovskites degrade when exposed to moisture or oxygen. (This makes perovskites the materials science equivalent of an MMA fighter who gets straight back up after a punch to the head, but is defeated by the tickling of a feather duster.) But because space is devoid of oxygen and moisture, Kirmani's lab has been developing materials already being tested in space. Funnily enough, his career in science began when, as a 19-year-old intern at the Indian Space Research Organisation, his modelling of solar radiation suggested a huge peak in the year 1859. Kirmani had independently stumbled across the Carrington Event. 'I had absolutely no idea at that time that I was actually going to be working on the materials side of this,' he says.

It would be gladdening to be able to conclude that the work of Kirmani and other scientists will make humanity immune to solar radiation. In the long run, however, Earth will eventually be blasted by radiation produced by a supernova – an exploding star – or by a gamma ray burst. Gamma ray bursts are more mysterious, but are probably produced by certain kinds of supernova and/or the collision of the collapsed cores of two supergiant stars. Toby Ord, a chronicler of existential risks, writes that supernova and gamma ray bursts would trigger potentially dangerous reactions in our atmosphere. Of these, 'the most important is probably the production of nitrogen oxides that would alter the Earth's climate and dramatically erode the ozone layer. This last effect is thought to be the most deadly, leaving us much more exposed to UV radiation for a period of years.'[25] To the relief of his readers, Ord quotes figures putting the likelihood of these interstellar blasts at one in several million per century.

These are problems for a superior iteration of humanity. So is the eventual expansion of the Sun, some five billion years from now, into a kind of star known as a red giant. This happens when a star's core finally runs out of hydrogen, triggering a round of extra-hot fusion involving helium. The Sun will expand far beyond its current size, engulfing Mercury, Venus and perhaps Earth. But if the predictions of futurists are correct, our species, or its successors, might one day have the ability to relocate our planet. Should intelligent life survive by that point, we can guess that it will have spread far beyond our Solar System and that its terrestrial, corporeal origins will be a distant memory. But perhaps that memory might be cherished in the same way that modern humans take care to preserve the sites of cave paintings. If our descendants choose to preserve the planet where it all began, they could send it further from the Sun by manipulating the orbit of asteroids or other planets[26] – thus executing a manoeuvre whose lineage goes back to the DART missions. Given time, Earth could be ferried even further from its original orbit.

For the time being, though, we are stuck in our current Solar System, hoping that John Kappenman is wrong about solar storms. It's a heavy burden that falls on your shoulders, I told him. 'I wish it would have been somebody more capable than me,' he said, chuckling. 'But, unfortunately, the world appears to be stuck with me for now.'

CHAPTER 3
STOPPERING A SUPERVOLCANO

'You might hear the shrieks of women, the screams of children, and the shouts of men ... the greater part convinced that there were now no gods at all, and that the final endless night of which we have heard had come upon the world.'

Pliny the Younger

Bring to your mind's eye the red-hot lava that belches forth from an erupting volcano. Down the slope it rolls, sizzling and smoking. This river of lava is far more representative of our planet's contents than anything else we see on the surface. Compared to the vastness of Earth's unfathomable depths of magma, this lava flow is about as significant as a drop of water is to all the world's oceans.

Most of the time, we are insulated from this roiling, superheated mass by many kilometres of solid rock. But when even a minuscule proportion of that magma emerges from Earth's crust, it is unignorable. Rivers of lava, plumes of ash, weeks of tremors: even an Enlightenment scientist could not ignore a volcano. Indeed, there is a long history of lunar craters being mistakenly attributed to volcanic eruptions rather than asteroid collisions.[1] Here on Earth, we have been assailed by such eruptions as those of Vesuvius in 79 AD,* Mount St Helens in 1980

* The quotation with which I began this chapter comes from Pliny's account of the eruption of Vesuvius.

and many in between. Likewise, Eyjafjallajökull in 2010 is easy to recall, if not to spell.

Those fearsome examples, though, are small fry. As with asteroid collisions, so with volcanoes: the vast majority of events are harmless, but a lesser-spotted minority might threaten our existence. In the case of volcanoes, that lesser-spotted minority is known as 'supervolcanoes'. There are two things you need to know about this category.

The first is that its members are gigantic. Scientists classify eruptions through a rough scale called the volcanic explosivity index (VEI), which refers to the amount of matter expelled. The VEI scale goes from 0 to 8, or 'effusive' to 'apocalyptic', with each level constituting a tenfold increase on the one below. VEI7 eruptions occur every 1,000 years or so, VEI8 eruptions roughly every 100,000 years.[2]

The lower the number, the more common the grade of eruption. But even the smaller eruptions are liable to cause chaos. In 2010, meltwater from the Eyjafjallajökull glacier, in the south of Iceland, mixed with hot magma, resulting in an explosive eruption that blasted unusually fine-grained ash into the air.[3] Fearing that this ash would cause jet engines to fail, airline managers cancelled more than 100,000 flights over the ensuing days.[4] Yet this volcanic event was a mere VEI3.[5] That means a supervolcano eruption is 100,000 times bigger than the Eyjafjallajökull eruption, and 1,000 times bigger than the 1980 eruption of St Helens in Washington state, the deadliest volcanic event in American history.

On which note: the second thing you need to need to know about supervolcanoes is that Yellowstone, the vast American caldera* whose heat sustains geysers and hot springs galore, is *not* about to erupt. 'I mean, you can't watch a documentary anymore without hearing that it's a "ticking time bomb" or a

* A caldera is a vast, bowl-shaped hollow that forms when a volcano's cone collapses, post-eruption, into the space left by the ejected magma.

"powder keg",' said Mike Poland, 'which is just a bunch of garbage.'

Mike Poland is the scientist-in-charge of the Yellowstone Volcano Observatory, so his assessment can probably be trusted. To ambulance-chasing scribes whose livelihoods depend on a non-trivial level of global imperilment, his verdict is a disappointing one. To anyone else, though, it should be a relief. The last time a supervolcano erupted on Earth, so much sulphur was belched into the atmosphere that, according to a high-profile analysis published in 2021, the planet cooled by several degrees and the ozone layer, which protects us from solar radiation, was severely depleted.[6] This 'volcanic winter', which is much like the 'impact winter' believed to be caused by large asteroids, devastates plant life by starving it of light. Herbivores are therefore starved of plants and predators are therefore starved of herbivores. That most recent supervolcano eruption, the one mentioned in the analysis from 2021, was that of Toba, which took place 74,000 years ago in what is now Indonesia. In 1998, the anthropologist Stanley Ambrose calculated that Toba reduced the world's human population to around 10,000 people.[7] This would constitute an exceptionally close shave for humanity, though Ambrose's analysis, as well as the occurrence of a Toba-induced volcanic winter, has been disputed.[8]

Supervolcanic eruptions – 'super-eruptions', to use the somewhat more sober academic term – are infrequent. In 2018, four researchers at the University of Bristol estimated that, in any given century, the chance of being destroyed by a super-eruption is about one in 170.[9] It is probably worth our monitoring the 20 or so supervolcanoes that exist worldwide.[10]

The work of Poland and his team is to keep an eye on Yellowstone, which is a well-loved American national park as well as an active supervolcano. Buffalo, wolves and bears populate its hills and plains; tourists plod around the perimeter of the many natural attractions animated by the heat of the subterranean magma. These attractions include the famous geyser Old Faithful, by which the author was spattered in 2023, and the brilliantly coloured Grand Prismatic Spring, whose deep-blue centre, green shallows and orange-red rim are the result of bacterial activity enabled by that subterranean heat. Poland and his colleagues use sensors to monitor what's going on underneath all this, keeping track of the vigour of the hydrothermal activity and the tremors that might herald danger. And for all the geological titillation administered to the public – the 'ticking time bomb' stuff about Yellowstone being on the cusp of eruption – there is, according to Poland, 'no sign of anything anytime soon'. There might be eruptions one day, he said, but they will probably be constituted of lava flows rather than a great big explosion. 'The magma chamber is mostly solid, and there's just no sign of anything brewing.'

More likely to occur in our lifetimes than a dangerous eruption, said Poland, is a big earthquake. Yellowstone sits on several fractures in Earth's crust, which results in frequent tremors. The region last suffered a serious earthquake in 1959, when 20 people were killed and a landslide threatened to cause a dam to be breached. Only some very quick digging by the US Army Corps of Engineers averted the flooding of an area of several square miles. Other dangers of such local earthquakes include the eruption of geysers and hot springs that have never before been recorded. And so, when politicians' aides visit Yellowstone, wanting to know how the US should be preparing for the eruption of a supervolcano, Poland tells them to prepare for earthquakes instead. It's possible, he says, that Yellowstone supervolcano might never erupt and that the North American tectonic plate might rumble over the hotspot, thus sealing it in to die.

And so Poland and his colleagues, having dealt with the aides, go back to their observations. The work strikes me as not dissimilar to watching a simmering pan of water. The water might never boil, but there is no way of turning down the heat.

Or is there? Some years ago, a group of scientists at NASA's Jet Propulsion Laboratory, one of the agency's research and development centres, came up with a proposal that I was confident Poland would hate. Then again, he was good-natured enough for me to guess that I could invoke it regardless. 'Mike,' I told him, 'I'm going to bring up a certain paper from the Jet Propulsion Laboratory and you're going to roll your eyes.'

I was correct. 'I've been dealing with this for six or seven years,' he said. 'It is *total* garbage.'

The paper has a more arresting title than those to which geologists are accustomed: 'Defending Human Civilization from Supervolcanic Eruptions'.[11] It was published in 2015, a few years before the DART mission, with the authors noting the 'considerable attention now focused on asteroid deflection techniques'. Given that the worst asteroid impacts now looked likely to be preventable, the authors proposed 'that corresponding efforts be considered towards supervolcanic eruptions'. Absent such efforts, they argued, humanity would have no defence against volcanic winter and the mass starvation that would follow.

To prevent this catastrophe, the scientists said, humanity would have to solve a single, but challenging, technical problem. Volcanic eruptions are the result of a build-up of thermal energy, so anyone seeking to prevent an eruption, they wrote, should find a way of wicking away a caldera's heat. One practical difficulty here, though, is that directly interfering with the magma might *cause* an eruption rather than prevent it. Another difficulty is that these volcanic systems harbour heat on a tremendous scale. Using conceivable methods, it would take tens of thousands of years to conduct away enough heat for the eruption to be averted.

The scientists, though, had noted exciting recent progress in the field of geothermal power. (In Chapter 7, I will explain these exciting advances in more detail. For now, let it suffice for us to note another instance in which progress in one line of anti-catastrophic endeavour has begotten progress in another.) In existing geothermal projects, engineers drill a pair of neighbouring holes before connecting them via a network of fractures. Thus, when you pump fluid down one pipe, it's hot by the time you draw it out of the other pipe. The NASA scientists, therefore, proposed that a few of these geothermal systems could be installed beside the Yellowstone magma chamber. One hundred and sixty of them, to be precise.

'Even for a massive supervolcano such as Yellowstone,' they wrote, 'it would take less than 50,000 years for such a cooling system to completely drain the heat away from the magma chamber, all the while generating electricity at competitive prices.'

It was a pleasingly neat solution to an extremely difficult problem. In 2017, the paper was featured in a BBC article entitled 'Nasa's Ambitious Plan to Save Earth From a Supervolcano'. That was when the public started asking Poland about the idea, and they still haven't stopped. 'Unfortunately,' said Poland, 'that particular report has become the stuff of legend. And I still get emails from people saying, "Well, have they started drilling it yet? They're doing their drilling right now, right? How far are they?"'

There has been no drilling. This is for several reasons, explained Poland. First of all, geothermal exploitation of Yellowstone is illegal. Of course, laws are sometimes repealed if there's a good reason to do so; the preservation of the human race would probably meet this threshold. But the scientists' plan, in Poland's view, had technical problems as well as legal ones. For instance, if magma is suddenly depressurised, it can – depending on what gases are mixed up in it – rise towards the surface. Poland has compared this to what happens when you put a can of soda in the freezer. Additionally, the project would require colossal amounts of water to be pumped long distances. Poland presumes this water would have to come from the Ogallala Aquifer, which yields about 30 per cent of the ground-water used in American irrigation.[12] 'That aquifer's kind of important to the whole Midwest,' said Poland. All in all, the plan 'would be a bit like trying to cool a pot of boiling water on the stove by dripping cold water into it, but not turning off the burner. It just wouldn't work.'

Poland is an engaging communicator, but for a moment he exuded a sense of don't-make-me-tap-the-sign weariness. He has been pestered about the proposal so many times that the United States Geological Survey website now bears an article, under Poland's name, titled 'Why Can't We Drill Yellowstone to Stop Eruptions and Make Power?'. And still the pestering continues.

As far as Poland is concerned, we are never going to stop volcanoes from erupting. The engineering challenge, he said, 'is just beyond description'. More promising, in his view, are attempts to anticipate eruptions and mitigate their effects. From orbit, satellites can detect heat emissions, ground deformations and release of gases such as sulphur dioxide, each of which can be harbingers of eruptions. On the ground, there are similar opportunities to learn from volcanic events. There are between 50 and 70 volcanic eruptions each year,[13] and each of these provides useful data. In many cases, volcanologists will fly in to gather data and to test equipment, such as drones and gas sampling kit.

For a period of about two decades in the second half of the twentieth century, the first people on the scene[14] were often a married couple: a pair of French volcano-botherers called Katia and Maurice Krafft.[15] These husband-and-wife volcanologists were not formally connected to any institutions, but they contributed reams of data to their field, as well as hours of high-quality footage.

'Clunking along in heat-deflecting Tin Man suits they'd designed themselves,' the essayist John Calderazzo put it, 'they'd approach a crater gloved hand in gloved hand, curly-haired and bearish Maurice scanning the sulphurous sky for whistling lava bombs, ferret-like, ninety-five-pound Katia studying the ground for orange cracks or rivulets of 2,000-degree F rock flows. Lava melted the soles of their boots, heat warped their camera lenses, awful-smelling gases could "process" the film in their video cameras long before it reached the lab.'[16]

The Kraffts made many contributions to volcanology, though their greatest claim to membership of the anti-catastrophe league was their contribution to a timely evacuation of the area surrounding Mount Pinatubo in the Philippines. It was early 1991 and small eruptions had resumed. As the weeks progressed, the ground began to tremble not only in the area around the volcano, but towards the top of its cone. Nevertheless, there was scepticism in the Filipino government that an evacuation was necessary. Evacuations are a fine art; they involve a politically inconvenient scale of disruption and are often based on guesswork rather than certainty. And the downside of a false alarm makes populations much more reluctant to obey evacuation orders in the future.

The Kraffts helped lead the creation of a video that used their own footage and that of others to demonstrate the lethality of eruptions. This video was shown to a series of government officials. The Kraffts and their team had agonised over whether to include dead bodies in the video, but it was the footage of a live girl, caught up in the carnage of a Colombian eruption in 1985, that is said to have caused audible gasps of horror. There she was, trapped and shivering in a cooled mudflow. Vulnerable children, it seemed, were even more persuasive than dead bodies.[17] The video is now regarded as having been instrumental in persuading the government to act. By 15 June, more than 60,000 people had evacuated the area that lay within a 30-kilometre (19-mile) radius of Pinatubo.[18] They thus escaped the worst effects of what was to become the second-largest eruption of the 20th century: a VEI6.[19]

Two weeks prior, the Kraffts had visited another volcano, Japan's Mount Unzen, to gather further footage of pyroclastic flows.[20] These are the fast-moving currents of gas and volcanic material that, at 1000°C, tear downhill at speeds of up to 450mph.[21] The hope of the Kraffts, and of their fellow volcanologist Harry Glicken, who had come with them, was that new footage would strengthen their warnings.

All three died. The remains of the Kraffts were found beside their rental car. Their last moments, it seems, were spent trying to film the pyroclastic flow that killed them. Their bodies were charred beyond recognition, identified only by their cameras and by Maurice's watch, stilled forever at 4.18 p.m. He had said he wanted to die 'at the edge of a volcano'.[22] He got his wish.*

Judging by footage of them speaking and working, the Kraffts were driven by complementary forms of obsessiveness. Katia was a devoted chronicler of geology. Maurice seemed to be drawn, in a primal way, to danger, sometimes talking of his ambition to paddle a reinforced canoe down a flow of lava. It is this kind of individual curiosity, particularly when it turns into obsession, that prompts humanity's first steps towards mastery of a great peril. This was true of the astronomers whose curiosity set in motion a chain of events that would end with DART hitting Dimorphos. It might also be true of Anders Sandberg.

Sandberg grew up in Hagalund, northern Stockholm, in the 1970s. 'It was mostly solidly boring Scandinavian welfare state,' he told me. As a boy, he was always interested in science and technology. An episode of bullying triggered in him, he recalled, the realisation that happiness is a choice. 'That set me on the path to becoming the cheerful guy I am.'

As a man, he is a genial polymath who was part of the University of Oxford's Future of Humanity Institute (FHI). This is an institute that will have a significant role in my story of the anti-catastrophe league. Normally, academic disciplines are

* In 2022, two films documented the lives of the Kraffts: *Fire of Love* and *The Fire Within: A Requiem for Katia and Maurice Krafft*.

divided by borders that sometimes seem as arbitrary as the straight lines with which European bureaucrats used to carve up foreign continents. But the staff of FHI, an institution that was putatively part of Oxford's philosophy department, were unconstrained by these boundaries.

Even among these multidisciplinarians, Sandberg stands out as an academic freewheeler. His research interests include, but are not limited to, global catastrophic risks; whole-brain emulation (a theoretical process by which a conscious mind could be digitised); the likelihood of intelligent life existing elsewhere in the observable universe; and the methods by which a highly advanced species might, billions of years from now, restructure the universe for their own benefit. More whimsically, he once wrote a paper that tried to answer the question: 'What if the entire Earth was instantaneously replaced with an equal volume of closely packed, but uncompressed blueberries?'[23]

Setting out the precise terms of the thought experiment, Sandberg elects for 'big, thick-skinned highbush blueberries' rather than the 'wild, thin-skinned blueberries' with which he was acquainted as a boy in Sweden. The bigger blueberries break less easily, but this blueberry Earth is nevertheless a hostile one.

Bang! The blueberry Earth is born fully formed. 'Standing on the blueberries might be possible in theory,' Sandberg writes, 'except that almost immediately they begin to compress rapidly and air starts erupting everywhere. The effect is basically the worst earthquake ever, and it keeps on going until everything has fallen 715 km. While this is going on everything heats up drastically until the entire environment is boiling jam and steam. The end result is a world that has a steam atmosphere covering an ocean of jam on top of warm blueberry granita.'

I've met Sandberg several times over the years, interviewing him on topics ranging from the deepest hole in the world* to

* An important topic, and one to which I shall return.

the wisdom, or otherwise, of contacting aliens.* Without fail, he is radiantly cheerful. If, in the process of sending Sandberg's brain into the cloud, you happened to run an analysis of its most unusual properties, you'd probably find an exceptionally high level of baseline happiness, as well as an exceptionally high level of curiosity.

And the prospect of Sandberg living long enough to have his brain uploaded is perhaps a little less remote than it is for the rest of us. Over his jumper sits a pendant that marks him as someone who has arranged to be cryogenically preserved after his death. People who have signed up for cryogenic preservation hope to be awoken at a time when pipe dreams such as whole-brain emulation have become reality. The pendants typically carry instructions – *do not embalm or autopsy!* – for the event that the wearer is found dead.

The pendant is a clue towards another unusual trait. Most of us are weighed down by the invisible bonds of social convention. We might be reluctant to sport such an unorthodox pendant, even if doing so might modestly raise our chances of eternal life. On Sandberg, though, those invisible bonds seem to weigh a little less heavy.

And this is an attitude that might be needed if humanity is to defuse supervolcanoes as well as deflect asteroids. As Mike Poland, the Yellowstone volcanologist, points out, there are legal and technical obstacles to overcoming the peril of supervolcanoes. There is also the obstacle of convention. The practice of interfering on a large scale with our planet's natural processes is known as 'geoengineering' and it is sufficiently unpopular that its status borders on taboo.[24] Environmentalists, for instance, loathe the idea of mitigating global warming by injecting sulphur dioxide into the atmosphere. (Perhaps this idea rings a bell; it is essentially a limited and self-inflicted volcanic winter.) They view this method, justifiably, as a failure to address the

* Ditto.

root causes of global warming, and a failure to address related problems such as air pollution. More profoundly, many people feel intuitively that planetary systems should be sacrosanct, that to meddle with them would be tantamount to 'playing God'.[25]

So far, only a very small number of people have dabbled in this dark art.* But although the world is not yet playing God, it is learning its lines. The US government, for instance, has funded research into solar geoengineering, via which humanity could reduce the world's temperature by injecting particular gases into the atmosphere.[26] With the world's global warming targets sliding further from view, the idea of intervening in this radical manner will become more attractive. As Sandberg once said to me of the climate: 'If you break it, you bought it.'

Similarly, he told me, we might one day want to protect ourselves by manipulating geological processes. 'Right now, we're mostly too weak as a species to affect the really big geological systems. But that was true for the climate, too.'

Sandberg became interested in this idea as a result of a chance meeting at a conference. In 2021, at an Effective Altruism conference in London, he was seated at a table with Lara Mani, a specialist in volcanic risk communication. Mani hails from the University of Cambridge's Centre for the Study of Existential Risk, which is a less kooky counterpart to FHI. Mani told the table about the Icelandic geothermal engineers who had accidentally drilled into magma chambers. The group was enraptured. Sandberg was desperate to know what happened next.

'I was almost disappointed,' he recalled, 'that the end result was a few lost drill bits. But, still, in our conversation, we quickly realised that this was going to happen. This is likely to be a relevant thing. And maybe you could defuse volcanoes in this way. Even if it's unlikely, some people are going to request it.'

* I will introduce one in Chapter 6.

Thus was born a cross-disciplinary partnership. Mani and Sandberg added Mike Cassidy, a volcanologist and petrologist* who, at this point, was at the University of Oxford and is now an associate professor at the University of Birmingham. Mani convinced Cassidy and Sandberg to join her in making a case for volcano geoengineering. Cassidy would bring his expertise in volcanology; Mani, herself a volcanologist, would add her specialism in volcanic risk communication. (You will recall the problems associated with evacuating people from the sites of potential eruptions.) Sandberg would focus on the ethics of the matter, as well as the historical precedents. Along the way, he said, 'I picked up enough geology to be dangerous and annoying'. (This new interest prompted Sandberg to start carrying around with him a dosimeter, a device that allows him to measure the radioactivity of rocks. To this day, he retains the dosimeter and the habit.)

The trio would go on to co-author a paper that is, to the best of my and their knowledge, the first academic examination of the ethics of volcano geoengineering. In the paper, the researchers gently point out that there are precedents for directly manipulating natural hazards, be they avalanches, forest fires, floods or landslides. (The DART mission gets a mention, too.) They also refer to the handful of attempts in the 20th century to redirect dangerous flows of lava. These redirections were well-intentioned, although there were some who proposed to use volcanoes offensively. Reginald Purbrick, a wartime British MP, was keen on the idea of bombing Vesuvius so as to trigger an eruption. Similarly, it was suggested to President Roosevelt that the same might be done to Mount Fuji in order to 'convince the mass of Japanese that their gods were angry with them'.[27]

Neither the British nor the Americans, fortunately, set a precedent for large-scale interference with volcanoes. Nevertheless,

* Petrology is the scientific study of the origin and structure of rocks.

argued the scientists, 'we believe there is a strong ethical case to support research into the efficacy and safety of volcano geoengineering for its potential future use'. After all, 'all of humanity has a common interest in the good of a stable geosphere': a world that is free of destructive eruptions and volcanic winter.

And we had better work out an ethical and regulatory framework soon, the scientists argued. Over the coming century, humans are likely to make increasing use of volcanoes, impelled by geopolitical, economical and environmental pressures to pursue geothermal energy, mineral resources, carbon sequestration* and climate change mitigation.† And although we might have many centuries before the next supereruption, it might also take many centuries to head off the threat. 'We may have duties to prevent harm to future people,' wrote the scientists, referring to the existing problem of the geological storage of radioactive waste. These moral duties, they said, 'occur because of our privileged early position in time and because some interventions like bleeding off magma chamber heat takes a long time'.

Volcanic engineering, therefore, 'should be regulated jointly for the benefit of all and future generations'. The scientists suggested some protocols and concluded that 'there is a strong ethical case to support research into the efficacy and safety of volcano geoengineering, even if we do not decide to conduct it'.

This is all predicated, of course, on the idea that large-scale, responsible volcano engineering is possible. Mike Poland, you'll remember, was sceptical. Where Sandberg differs is that he expects human technological capability to dramatically improve. Harvesting large amounts of thermal energy, he told me in 2024, will probably become more straightforward as our civilisation matures. (And given the power-hungry nature of

* Volcanic ash, scientists at the University of Southampton found in 2020, is good at absorbing atmospheric carbon dioxide.[28]

† It has been noted in some quarters that the release of volcanic aerosols might help defray global warming. File under 'Do Not Try At Home'.[29]

our existing and forthcoming AI systems, we will be quick to find uses for this energy.) 'We tend to forget,' Sandberg said, 'how unreasonable it is that we can do the things we already do. I flew in from Miami, briefly dropped by Oxford and now I'm in Stockholm. There's nothing to it except jetlag. Just 100 years ago, that would have been a major achievement. These unthinkable things become more thinkable over time.' (I'm reminded, again, of the DART mission, which took place little more than a century after the Wright brothers' first demonstration of human flight.)

To say these things can't be done, argued Sandberg, is to miss the point. Instead, he says, we should ask: 'When can it be done? And what are the preconditions for doing it safely and sanely?'

One can imagine this capability coming in handy some time before the next super-eruption. Even a relatively moderate eruption can cause worldwide harm. Take the eruption, in 1815, of Mount Tambora. Though this VEI7 event took place in what is now Indonesia, it threw so much sunlight-dimming soot into the air that there followed crop failure in western Europe and the north-east of the United States.[30] For some time, this 'year without a summer' was romantically suggested to have driven Mary Shelley indoors to write *Frankenstein*. Alas: killjoy academics have since sought to dismantle this myth.[31,32]

More troublingly, several active volcanoes currently menace key economic choke-points. Take the Malacca Strait, which links the Indian Ocean to the Pacific. The cargo of ships using the strait – grain, crude oil, and every other commodity under the Sun – comprises an estimated 40 per cent of global trade.[33] Above these ships is one of the busiest air routes in the world, while below them, running along the seabed, is a dense array of submarine communications cables that carry a significant proportion of global internet traffic.

Together, these factors make the Malacca Strait one of the most vital arteries of the global economy. Every year, approximately 90,000 ships pass through this slender sea lane.[34]

As such, the strait has been classified as a trade chokepoint in reports by the World Trade Organisation, the US Energy Information Administration and Chatham House, the London-based foreign affairs think tank.[35,36,37] All of which is to say: nice strait you've got there. Be a shame if something ... happened to it.

The greatest menaces to the Malacca Strait, which separates the Malay Peninsula from the Indonesian island of Sumatra, lie in the natural world. Of the many intriguing maps of activity in the area, the most arresting is the one that collates the world's active volcanoes and recent earthquakes.* As Mani pointed out in 2021, in an article published by *Nature*, the region is 'highly volcanically active'.[38] On Sumatra, the island to the south of Malacca, lies Mount Sinabung, which spewed volcanic materials 5km into the air in 2021.[39] Also on Sumatra is Toba, the dormant supervolcano with which humanity has previous. And on the island of Java, which is some way south-east, lies Mount Merapi, which was producing lava flows at the time of writing (July 2024).[40]

And an eruption, as Mani told me, 'can still really disrupt the strait. And the thing is, when a volcano starts, it doesn't tell you when it's going to stop, or how big the eruption is going to be.'

It's worth imagining what would happen if one of those active volcanoes produced an eruption that would qualify as a VEI5 or VEI6. (Globally, there is a one-in-10 chance each year of there being a VEI5 eruption, and a one-in-100 chance, approximately, of a VEI6.) Let's imagine the volcano in question is Merapi. Tremors shake local towns, ash fills the sky and magma bursts forth from Merapi's cone.

Much depends on the direction of the wind. If humanity is lucky, the wind is north-easterly and the ash clouds are swept out over the empty expanse of the Indian Ocean. If humanity is less fortunate, however, the wind is south-westerly. All air traffic in the Malacca Strait is grounded. The ash falls onto the

* The curious can find this website at volcanoesandearthquakes.com.

strait itself. On the surface of the sea, rafts of pumice accumulate.

Don't forget those undersea internet cables. If the eruption, said Mani, 'was close enough to the coast, and it could actually interact with the strait itself, you would get something like turbidity currents'. These, she explained, are clouds of fast-moving material that can travel over and under the water. 'And that's typically what severs cables.' If those cables are severed, regional internet blackouts would be likely to follow. And regional internet blackouts would cut off Singapore from the rest of the world, causing disarray in the financial markets.

In the worst case, an eruption would be accompanied by a large earthquake. This would not just cause turbidity currents. It could also cause a tsunami to hit the strait, as the Boxing Day tsunami did in 2004.[41] This would cause human devastation on an immense scale. One hopes that the Singaporean laboratory that handles some of the world's deadliest pathogens is prepared for flooding.[42]

Surprisingly enough, the shipping industry might emerge relatively unscathed. Earthquakes and eruptions would cause less disruption to global shipping, said Tristan Smith, of University College London, than did the blocking of the Suez Canal, in 2021, by the container ship *Ever Given*. Smith is a shipping expert at his university's energy institute. He told me that machinery should be able to cope with ash and that a tsunami is more dangerous to people on land, where the wave breaks and is at its biggest, than on sea.

Smith forecast that, in the case of an eruption, an exclusion zone would be declared, forcing ships to take a different route. Naturally, the rerouting of ships would incur delays and financial cost. Ultimately, though, the system should be able to handle it. Smith said: 'If you have a ship that gets held up by three days, because it has to go the long way around Indonesia, all that ship needs to do is increase its speed by one or two knots and that delay is counteracted.'

It is no coincidence that important human activity often seems to be in the neighbourhood of volcanoes. Eruptions bring minerals to Earth's surface, making volcanic islands highly fertile. Volcanic regions are well-endowed with natural harbours, making them promising sites for settlement and trade.

We are unlikely to abandon these sites in the foreseeable future, but we can still make them safer. And, doing so, we must find ways of lessening the fragility of choke points such as the strait of Malacca. Our global networks tend to concentrate rather than decentralise, which creates vulnerability. Scenarios like those of Merapi and Malacca are a good illustration of the phenomenon of 'cascading risks': hazards that ripple through global systems, causing outsized damage.

In the long term, we might be able to insulate ourselves by defusing volcanoes. For now, we must improve our ability to monitor them. As Mani told me, Indonesia has 'more volcanoes than you can shake a stick at'. Many of them have not been studied in detail, a situation typical of global volcanology. Only around 30 per cent of the world's active volcanoes are monitored, Mani told me.

As volcanologists often point out, their field is funded much less generously than is asteroid-hunting. In a more resilient world, volcanologists would not have to 'piggyback', as Mike Poland put it, off existing satellites, but would have exclusive use of several small satellites of their own.

Elsewhere, the best preparation is diversification. Satellites carry a small fraction of internet traffic. The more satellites we use to connect ourselves to the internet, the lesser our dependence on undersea cables. (Naturally, those satellites ought to be insulated against solar storms.) This will also reduce our vulnerability to deliberate attacks on those undersea cables. In March 2024, Yemen's Houthi rebels cut three cables under the Red Sea.[43] The Red Sea, of course, is another consequential choke point.

And a milder form of geoengineering might also be of use. Thailand wants to create an overground bypass that cuts out the circuitous Malacca Strait route.[44] More ambitious, though, is the as-yet-unrealised Chinese plan to construct a canal across southern Thailand.[45] The purpose of this is to shore up China's geopolitical resilience, but a welcome side-effect would be a lessening of global trade's reliance on the Malacca route.

All in all, serious volcanic events are low in probability, at least on a century-by-century basis, but severe in their potential impact. And some time in the very far future, we might face volcanic hazards even worse than supervolcanoes. Take the Siberian Traps. This large region of volcanic rock is the legacy of a vast field of lava, seven million square kilometres (three million square miles) in area. For two million years, near-continuous eruptions would have made this enormous area a hell on Earth. This was the period in Earth's history in which occurred an extinction event so calamitous that it is known as the 'Great Dying'.[46] The Siberian Traps event seems to have taken place around 250 million years ago and the risk per century of another such event was put at one-in-200,000 by the extinction risk researcher Toby Ord. This is probably one for our descendants to worry about, in which case we must hope that Sandberg is correct in his forecast that they will be more capable than we can possibly imagine.

Mani is down-to-earth and pragmatic. In 2022, I asked her about the burden, if that's what it is, of warning the world about volcanic eruptions. 'We certainly shouldn't be afraid of risks like this,' she said, 'and they shouldn't keep people awake at night.' She said she hoped that her work 'makes our societies want more accountability, and to know their governments are doing all they can to prepare for high-impact risk events'.

As such, she is not given to doomerism. 'You have to be optimistic about humanity's future to work on existential risks, and I certainly am.'

Another irony of volcanism, and an in-joke among researchers, is that it often takes place in scenic locations. With the Siberian Traps quiet, it is places such as the 'Pacific Ring of Fire', which encompasses thousands of kilometres of coastline and 90 per cent of the world's volcanoes, that attract academic attention. Sunburns, in most of these places, are generally more likely than first-degree burns. One can imagine our descendants eliminating both.

CHAPTER 4

'WE WILL DIE, BUT WE WILL SINK THEM ALL!'

'The falcon cannot hear the falconer'

W. B. Yeats

Up to this point, the hazards I have discussed have generally come from above us (asteroids, solar storms) or below (super-eruptions). These horrors are known as the natural risks – those that come from nature. If civilisation were a household, the natural risks would be reason enough to fix the roof, strengthen the shutters and install a lava-proof lining in the basement.

But, as any student of horror knows, it is sometimes the case that the call is coming from inside the house. In the annals of catastrophic risk, the 20th century is a significant one. That is not because of our exploits with asteroids, nor because of our waking up to the threat of solar storms and super-eruptions. Rather, it is because the 20th century was the century in which we invented a catastrophe of our own: nuclear holocaust.

It is a peril that needs little introduction, particularly to readers who lived through the Cold War. My father, who grew up in Northamptonshire in the 1950s and early 1960s, lived near an airbase, RAF Wittering. The UK had joined the USA and the Soviet Union in acquiring a nuclear arsenal, and the weapons were, in this era, to be carried in V-bombers. These were enormous long-range jet bombers equipped to drop nuclear weapons on Moscow itself. The bombers were not hidden from the public, as modern nuclear-armed submarines are. Instead, they were sometimes visible; my father remembers seeing them. How disquieting it must have been to see them taxiing on the runway during the excruciating peak of the Cuban Missile Crisis in 1962. Had the V-bombers been sent to drop even a single nuclear bomb on Moscow, it would have been either a response to a Russian attack or the trigger for one. It would have heralded a spree of destruction and suffering unparalleled in human history – the making of a hell on Earth.

Hell on Earth, in this respect, is the product of chain reactions. The first takes place among certain isotopes of plutonium or uranium.* Human insight into this phenomenon began slowly, then accelerated – an intellectual chain reaction to rival the sub-atomic chain reaction. It was in 1911 that the physicist Ernest Rutherford, a New Zealander then chairing the physics department at the University of Manchester, proposed that the atom had a nuclear structure: electrons orbiting around a heavier core, the nucleus. Three years later, war broke out. One of Rutherford's students, James Chadwick, was held by the Germans in a prisoner-of-war camp. While captive, he set up a rudimentary laboratory in which he used limited materials, such as radioactive toothpaste, to conduct experiments in atomic physics. Once released, Chadwick returned to academic physics, resuming his work under Rutherford. In 1932, now at the University of Cambridge, Chadwick proved the existence of the elementary particle we know as the neutron. Neutrons have no electrical charge, which makes them hard to detect. But it also makes them the right sort of missile to be fired at a nucleus, whose charge would usually repel an oncoming particle.

* One isotope of an element differs from another by having more or fewer neutrons in its nucleus. 'Heavier' isotopes have more neutrons. Heavier isotopes tend to be less stable and more inclined to undergo radioactive decay.

Here began the acceleration of humanity's understanding of nuclear fission. Weeks after Chadwick discovered the neutron, two of his colleagues at Cambridge bombarded lithium atoms with high-speed protons. These colleagues were John Cockcroft and Ernest Walton, themselves charges of Ernest Rutherford. Sitting in a lead-lined box on 14 April 1932, Walton watched a screen that was designed to flash when hit by helium. Whereas lithium atoms are composed of four neutrons and four protons, helium particles are made of two apiece. Cockcroft and Walton had little expectation that their screen would flash on the first attempt, not least because their particle accelerator was made of such high-tech materials as children's plasticine.*[1]

And yet the screen flashed. Walton fetched Cockcroft and Rutherford. The protons had made helium atoms of the lithium atoms, meaning that this was the first time that humanity had split an atom.

Protons cannot sustain the kind of chain reaction that releases huge quantities of energy, and neither can lithium. But before the decade was out, two German chemists† demonstrated that by bombarding uranium with neutrons, they had produced an element called barium.[2] This was another splitting of the atom, but a more consequential one. In his theory of relativity, Einstein had made the implicit prediction that matter could be turned into energy, and he was now being proven correct. Whereas the Cambridge physicists had released only a small amount of energy, the German chemists released much more. In firing neutrons, the particle discovered by James Chadwick, they had used the right sort of missile. And in firing neutrons at uranium, they had used the right sort of target. Being one of the heaviest elements, uranium is an ungainly mass of particles. This makes it prone to nuclear reactions.

* This versatile substance, I hasten to clarify, was deployed as a sealant rather than as a structural material.

† Otto Hahn and Fritz Strassmann, in 1938.

Most scientists did not realise the significance of these advances. But to those in the know, the arrival of the latest edition of certain scientific journals must have been as exciting as a new model of ChatGPT is today. Then, as now, beneath the fanfare of progress could be heard the drumbeat of worry. Leo Szilard, a Hungarian physicist who had just fled Europe for the USA, guessed that high-energy chain reactions were not only possible, but scientifically imminent. On the one hand, such a chain reaction might provide humanity with an awe-inspiring new source of energy; on the other, it could be a tool of immense destruction. (Rutherford had derided the idea of harnessing atomic energy as 'moonshine'.) But further experimentation, some of it led by Szilard, made the prospect of a destructive chain reaction even more real. In August 1939, with another world war lurching into reality, Szilard wrote a letter to which he persuaded Einstein to give his signature.[3] The letter was addressed to President Roosevelt and it warned him that uranium could, in the immediate future, be turned into not only 'an important source of energy', but also 'extremely powerful bombs of a new type'.[4] In Europe, the letter noted, uranium had been discovered in some mines in Czechoslovakia. Having just annexed the country, Germany had banned the mines' operators from selling the uranium.

The Einstein–Szilard letter was delivered in October 1939, by which point another world war had indeed broken out. On the other side of the Atlantic, James Chadwick and John Walton were among a group of British scientists who, in secret, conducted research into the feasibility of a uranium bomb. 'We entered the project with more skepticism than belief,' the scientists wrote in their report, 'though we felt it was a matter which had to be investigated. As we proceeded we became more and more convinced that release of atomic energy on a large scale is possible and that conditions can be chosen which would make it a very powerful weapon of war.'[5] The Brits, like Szilard, feared that the Germans might beat the Allies to the creation of that weapon. They shared their findings with the Roosevelt administration. Thus, in 1941, began what became known as the Manhattan Project. The race was on.

It was a race that, in some striking ways, prefigures the current race towards super-intelligent AI. There was isolated progress, largely ignored, towards an apparently faraway technology. Then, things began to move faster, both in the democratic world and elsewhere. All of a sudden, the scientists who pioneered the technology began issuing public warnings about its dangers. (As a loose analogue to Einstein and Szilard, take Geoffrey Hinton, known as the 'godfather of AI'. In 2023, Hinton quit his job at Google.[6] He has since publicly urged governments to take immediate action to ensure that AI is developed safely.) Just as the USA and Germany competed to be first to build the bomb in the 1930s, so the USA and China are currently competing to build super-intelligent AI in the 2020s. And just as the supply of uranium was a key constraint in the 1930s, so semiconductor chips are a key constraint in the 2020s. Secrecy is precluding cooperation, and the balance of power is liable to tilt decisively and destructively.

The outcome of the AI race is yet to be decided, but I doubt I need to remind readers who won the race to build the bomb. In 1942, the Manhattan Project repurposed a University of Chicago squash court to build the world's first working nuclear reactor from 'a crude pile of black bricks and wooden timbers'.[7] Those were the words of Enrico Fermi, who oversaw the experiment. Fermi's work in Europe was part of the series of breakthroughs that preceded the Manhattan Project. In 1938, he had won the Nobel Prize in Physics for his work on the artificial radioactivity produced by neutrons and on the nuclear reactions brought about by slow neutrons. Having fled Europe with his Jewish wife, Fermi was appointed an associate director of the laboratory at Los Alamos that conducted the Manhattan Project. The director was Robert Oppenheimer.

One physicist, Joseph Rotblat, walked away from the project. Rotblat believed that the Germans would be overpowered before they were able to make a bomb and that the technology should therefore not be brought into the world. But the project went ahead all the same. Two years and seven months after the experiment on that Chicago squash court, Oppenheimer was in charge of the Trinity test: the first attempt to detonate a nuclear weapon. The test took place in a New Mexico desert known as the Jornado del Muerto: or, in English, 'the route of the dead man'. At 5.29 a.m., on 16 July 1945, a physicist called Kenneth Bainbridge remotely detonated a device called 'the gadget'. Several miles away, a burst of neutrons collided with unstable plutonium that had been made from decayed uranium. Watching from afar was a small group of VIPs, including Oppenheimer and James Chadwick. What they beheld, at the moment of Bainbridge's detonation, was the world's first use of a nuclear bomb.

For a brief period, there was a point just above the surface of Earth that was ten times hotter than the surface of the Sun.[8] There was the world's first blinding flash of light from a nuclear bomb, the world's first nuclear shockwave, the world's first mushroom cloud. Major General Thomas Farrell, second-in-command to the general who directed the Manhattan Project, wrote an arresting account of what he saw. Reading it, one is reminded of the testimonies of those who saw the Carrington Event encolour southern skies. But the beauty Farrell describes is a terrible one. 'The effects,' he wrote, 'could well be called unprecedented, magnificent, beautiful, stupendous, and terrifying. The lighting effects beggared description. The whole country was lighted by a searing light with the intensity many times that of the midday sun. It was golden, purple, violet, gray, and blue. It lighted every peak, crevasse and ridge of the nearby mountain range with a clarity and beauty that cannot be described but must be seen to be imagined.'[9]

Oppenheimer, famously, recalled some lines from the *Bhagavad Gita*. 'If the radiance of a thousand suns were to burst at once into the sky, that would be like the splendor of the

Mighty One ... I am become death, the shatterer of worlds.'*[10] Bainbridge, more prosaically, remarked: 'Now we are all sons of bitches.'

Ten days later, the Los Alamos team sent to the island of Tinian the components of a ten-foot, 4.4-ton bomb.[11] Tinian was in the western Pacific and had been captured from Japan in 1944. Since then, it had become the main airfield for Japan-bound bombing missions. Even Tinian's commanding officer, though, did not know what was now being assembled in an air-conditioned shed. At 3 a.m. on 6 August, three weeks after the Trinity test, a modified B-29 bomber took off from one of Tinian's main runways. On board were twelve men and one Little Boy.[12] That, of course, was the name of the device in the aircraft's bomb bay. It was the younger brother of the Gadget and it was a device that was simpler and heavier. Instead of plutonium, which allows engineers to miniaturise their designs, Little Boy used only uranium. This was a cruder construction than the Gadget, but a reliable one. So heavy was its deadly cargo that the B-29 could barely take off.

Its flight was several hours long. En route, the crew's two weaponeers visited the bomb bay to insert four silk bags of cordite, an explosive, into Little Boy. The cordite would later ignite the chain reaction. When the plane drew closer to the Japanese coast, the chief weaponeer clambered back into the bomb bay to swap the Little Boy's green safety plugs, which prevented accidental activation, for red ones, which connected the electrical circuitry.[13] The bomb was now armed and, at 8.16 a.m., it was dropped, from high altitude, over Hiroshima. As the bomb fell, the air crew took a sharp turn and accelerated away.

Inside the bomb, the mechanical countdown had begun. About 43 seconds after it had been dropped, and at an altitude of 1,900 feet, Little Boy detonated its internal stock of cordite. The detonation rammed a projectile of uranium-235 into a

* Often rendered 'the destroyer of worlds'.

larger cargo of uranium. Neutrons pounded unstable atoms and the chain reaction took place in less than a millionth of a second. The fleeing B-29 was now miles away, but it was still buffeted by the shockwaves.

It was a sunny Monday morning. At 8.15 a.m., Hiroshima had been a city of about 300,000 people, with commuters cycling to work and soldiers doing their morning workouts.[14] Moments later, a fireball 370m (1,200 feet) wide turned those within it into ash. A blast wave levelled two-thirds of the city. Some of the bomb's victims were heated to such high temperatures that their only remains were shadow-like indentations on the ground. Some 70,000 people are believed to have been killed instantly,* and many more were horrendously injured. One survivor, a 13-year-old girl called Setsuko Nakamura, later recalled a ruined city whose air was so full of soot that it seemed night had fallen. At first, she thought her fellow survivors were ghosts. 'Their skin and flesh were hanging,' she would later recount, 'and parts of the bodies were missing. Some were carrying their own eyeballs.'[15]

Many of those who survived the initial effects of the bomb were so badly irradiated that they suffered similarly gruesome fates. The bomb blast had included a tsunami of gamma rays, the highest-energy form of electromagnetic wave in the universe. We do not feel gamma rays when they pass through us, but we feel their effects. The radiation damages our cells and DNA, resulting in symptoms such as vomiting, diarrhoea, fever, loss of head hair, bleeding under the skin and bleeding from the eyes. As a result of their radiation poisoning, many more Hiroshima residents died within weeks. Some developed bone marrow disease, where one's bone marrow fails to develop enough white blood cells to fend off infections, or cancer.[16]

Little Boy had carried 64 kilograms of enriched uranium, of which a mere 0.7 grams – the weight of a five-pound note – had

* Estimates vary significantly. 70,000 is the lower bound.[17]

been turned into energy. But it unleashed a blast that, as President Truman told the American people in a broadcast later that day, was more than two thousand times as powerful as the next-biggest bomb ever used in warfare.[18]* Japan, whose wartime culture demanded tenacity to the death, and whose leaders seem not to have immediately understood the scale of the devastation in Hiroshima,[20] refused to surrender. Two days later, the Soviet Union declared war on the Japanese. And a day after that, on 9 August, the US dropped another atomic bomb – this time on the city of Nagasaki. Tens of thousands more lives were lost, another city turned into wasteland. Historians continue to debate the relative importance to Japan's surrender of the atomic bomb detonations and the declaration of war by the Soviets, but a surrender it was. The Second World War was effectively over – settled by the winners of the deadliest arms race ever seen.

The bombing of Hiroshima and Nagasaki comprised the most acute loss of human life in history. These events were catastrophes. That much is clear from the broadcast by which the emperor of Japan notified his people of his decision to surrender. It was the first time his voice had been heard on the radio.[21] 'The war situation has developed not necessarily to Japan's advantage,' he said, explaining that 'the enemy has begun to employ a new and most cruel bomb, the power of which to damage is indeed incalculable'. The traditional telling is that the bomb ended a war that would otherwise have taken even more lives; modern historians have it that the brutality of the incoming Red Army weighed even heavier on the minds of the Japanese.[22]

Joseph Rotblat, the conscientious objector who then dedicated his life to nuclear disarmament, won a Nobel Peace Prize

* For those who are wondering, this was the British 'Grand Slam'. Forty-two bombs of this kind were dropped on Germany in March and April 1945.[19]

in 1995. But was there heroism to the work of Oppenheimer, Szilard, Chadwick *et al*? It is an uncomfortable question that yields no easy answers. Professor Lord Martin Rees, who is the astronomer royal as well as a specialist in the study of existential risk, knew several veterans of Los Alamos, including Rotblat. 'Those uprooted from placid academic labs to join the Manhattan project,' Lord Rees wrote in 2003, 'were deeply marked by their involvement.'[23]

Klaus Fuchs was another German scientist who fled the Nazis, arriving in England in September 1933. As part of a British delegation sent to participate in the Manhattan Project, Fuchs helped his colleagues work out how best to enrich uranium. But he was passing on the information to Soviet intelligence agencies, providing such a useful service that he is believed to have accelerated the USSR's atomic bomb programme by as much as two years. Fuchs was arrested in January 1950, but by then the Soviets had already conducted their first successful nuclear test.[24]

The UK became a nuclear weapon state in 1952 and led the world* in its construction of nuclear power stations.[25] These power stations, like their cousins either side of the Iron Curtain, were a bountiful dividend of the world's investment in nuclear science. Nuclear power stations began to proliferate, but so did nuclear weapons. France and China soon joined the nuclear weapons club, in 1960 and 1964 respectively,† though the two most powerful nuclear states, by far, were the USA and the USSR. By the early 1960s, the period in which my father saw the British V-bombers, the USA had 25,000 nuclear warheads. Russia started slower, and was poorer – but determined. It made

* Yes, there was such a time. The only regard in which Britain's nuclear power facilities currently lead the world is in their extraordinary cost.

† Israel, too, is thought to have joined the club in the 1960s, though this has never been officially acknowledged.[26]

tens of thousands of warheads, continuing until it had, by some distance, the world's largest nuclear arsenal.[27] Another chain reaction had reached a perilous zenith.

Having narrowed down our study of destruction to the single unit of a Hiroshima bombing, we must now widen our scope again. The atom bomb had been surpassed by a new weapon, one in which the initial fission reaction is used to start an even more powerful fusion reaction. Nuclear fusion, you will recall from earlier in this book, is the process that powers the Sun. Now it was powering bombs that were hundreds or thousands of times more potent than the bombs that levelled Hiroshima and Nagasaki.[28] Because the explosive power of these weapons results largely from the fusion of hydrogen nuclei, the bombs became known to the public as 'hydrogen bombs'. Scientists called them 'thermonuclear bombs', or, initially, 'super bombs'. The prospect of their development marked a Rubicon that even Oppenheimer could not ford. In 1949, he chaired a committee that assessed the merits of the US government's plan to build a hydrogen bomb. The group's conclusion makes an arresting contrast with the Szilard letter that helped spark the Manhattan Project. 'The extreme dangers to mankind inherent in the proposal wholly outweigh any military advantage that could come from this development,' Oppenheimer and colleagues wrote. 'Let it be clearly realized that this is a super weapon; it is in a totally different category from an atomic bomb.'[29] The weapon, they said, could be used as a tool of genocide. 'We believe a super bomb should never be produced.'

And yet the super bombs were produced, by the tens of thousands and by multiple governments. They were affixed to delivery devices whose speed and versatility far outstripped that of the B-29s that bombed Hiroshima and Nagasaki. Those planes were propeller-driven and could have been straightforwardly shot down before igniting the bomb, but the new generation of warheads could be dropped from jet planes; the UK's V-bombers were aircraft of this variety. Soon the bombs could be delivered via intercontinental ballistic missiles, known for short as ICBMs. These weapons are to be fired out of secret silos before arcing into space and back down towards a city at tremendous speed. ICBMs are hypersonic, which means that they can travel at more than five times the speed of sound, and they cannot be recalled once dispatched. Not much later, both the Western and Communist powers developed missiles that could be dispatched from nuclear-powered submarines. Equipped with these submarine-launched ballistic missiles (SLBMs), a vessel can surface near an enemy shoreline to inflict an unstoppable surprise attack. ICBMs and SLBMs are so hard to defend against that their development was probably more significant than even the creation of the fusion bomb.

Knowing themselves to be vulnerable, each side redoubled its ability to intimidate.[30] Both the Americans and the Soviets adopted military doctrines requiring them to fire back once they had detected an incoming missile. And firing back would involve all-out war: a pre-emptive levelling of missile silos, military bases and major cities. In 1960, the Pentagon's analysis of its own war plan found that a pre-emptive American first strike would kill a fifth of Earth's population. That figure, baleful though it was, did not take into account the Russian reply. You can detect an ICBM coming, after all; with 20 minutes' notice, you don't have time to evacuate a city, but you *do* have time to respond in kind. And if you don't make it plain beforehand that that's how you might respond, then your enemy can bully you as violently as he likes. The possibility of posthumous vengeance is therefore a central component of nuclear deterrence.

The era of plasticine sealant and repurposed squash courts was long gone, yet the systems that surrounded these deadly new weapons were nevertheless terrifyingly rickety. In 1961, while soaring above Goldsboro, North Carolina, an American B-52 bomber, manufactured by Boeing, began to collapse in mid-air. The plane's crew abandoned the aircraft (three of them died). The B-52's payload included two hydrogen bombs, of which one was equipped with a working parachute mechanism that caused it to flutter back down to Earth. The other hydrogen bomb slammed into a muddy field at more than 700 miles per hour.[31] Neither bomb detonated, but the mishap remains flabbergasting. Today's beleaguered Boeing executives, who had to explain in 2024 why a door came off one of their planes mid-flight, can at least protest that the company has had worse days at the office.

It is unlikely that a detonation, in the North Carolina incident, would have led to nuclear armageddon. Instead, the most dangerous moments of the Cold War were when hair-triggers were all but triggered – even erroneously. On 14 October 1962, American reconnaissance, conducted via a U-2 spy plane that could take high-quality images from far above the ground, confirmed what analysts had feared: that Cuba, the US's nearest Communist neighbour, was now constructing Soviet missile bases.[32] This panicked the Americans and marked the beginning of the Cuban Missile Crisis. Believing that the missiles had yet to arrive from the Soviet Union, President Kennedy's advisors told him to bomb the missile sites and invade Cuba. Instead, Kennedy imposed a blockade on 22 October. Thus began a stand-off so perilous that the US's defence readiness condition, DEFCON, was moved to level two – one step from nuclear war. Recollections from those 13 days suggest an atmosphere of terror.[33] Robert R. Johnson, author of a book on nuclear history, recounts being in school when a teacher interrupted a lesson to turn on minute-by-minute updates from the radio. 'One young girl cried out, "I will never see my mother again!" Another vomited.' A US air base official later recalled being told to say goodbye to his spouse; an 11-year-old girl from Ohio, Rose Gottemoeller, simply had 'a vague anxiety that something bad could happen'. (Remember her name.)

Although the public was unaware, the most perilous moment of all occurred on 27 October. A Soviet submarine, B-59, had been cornered by 15 American destroyers. With the submarine at their mercy, the American crews appeared to have the upper hand, but were ignorant of two crucial facts. One was that the Soviet submarine was equipped with a nuclear-tipped torpedo, able to cripple entire groups of vessels at a time. The other was that the submarine's crew was out of contact with Moscow. This meant that the exhausted Soviet submariners, who had been suffering the effects of an overheated vessel with high levels of carbon dioxide, were authorised to make unilateral use of their torpedo. If war appeared to have broken out, the crew could fire the weapon – turning the conflict nuclear. B-59 was surrounded by US vessels that were dropping depth charges, which are underwater bombs. 'It felt,' a crew member would later write, 'like you were sitting in a metal barrel which somebody is constantly blasting with a sledgehammer.'[34]

The depth charges were intended to signal to the B-59 crew that they were required to surface, but it was an inarticulate way of saying so. The submarine's captain, convinced he and his men were under attack, shouted for the torpedo to be prepared for launch. 'We're going to blast them now!', he is said to have roared. 'We will die, but we will sink them all!'[35]

The torpedo could not be launched without the agreement of the second- and third-most senior officers. Vasili Arkhipov, the second captain, vetoed the captain's panicked proposal.[36] We will never quite know what happened aboard that submarine, but we can count ourselves fortunate that the torpedo never emerged. Rattled by the crisis, the Americans and the Soviets began the diplomatic process that would eventually lead to nuclear weapons treaties. They also put in place a communications system that would ensure their leaders could always communicate. This system was known as the 'Hotline' – or, more evocatively, the 'Red Telephone'.

Given how many thousands of nuclear warheads were made, it is unsurprising that the Arkhipov incident was not the closest call. Two decades later, when warhead numbers were at their peak and Russo-American relations were at another nadir, a

Soviet lieutenant colonel was settling in for his overnight shift at a bunker near Moscow. It was 26 September 1983 and Stanislav Petrov was in charge of monitoring the system that warned the Soviet Union of incoming missiles. Shortly after midnight, alarm bells began to sound. An American ICBM, the system reported, was on its way. It was Petrov's responsibility to pass this information up the command chain. Those at the top would have a matter of minutes to decide what to do.

But Petrov guessed that an American attack would involve more than one missile, so he waited.[37] The siren howled, and his back-lit red screen displayed the word 'Launch', but he kept waiting. Then the computer notified him of a second missile, then a third, then a fourth. On the phone, another officer was shouting at him to do his job.

All Petrov had to do, he later recalled, was to ring the most senior commanders. He had a funny feeling in his gut, though, a feeling that this was a false alarm.[38] There should have been more missiles, and there should have been more data; elsewhere, radar installations had not picked up incoming missiles. As many as 200 people were watching him, but Petrov had made his decision – and he stuck to it.[39] In fewer than 20 minutes, either the missiles would hit or he would be vindicated. 'I felt,' he said, 'like I was sitting on a hot frying pan.'[40]

Twenty agonising minutes later, Petrov was vindicated rather than dead. Had he followed protocol, there is a good chance that the USSR would have sent thermonuclear bombs screaming towards the West, raining hellfire on major cities and killing hundreds of millions of people. The West would have responded in kind. In the ashen wastes that remained, millions more would have died of radiation poisoning. Those that survived the radiation would have faced starvation; the major Western powers would have little left in the way of government, let alone infrastructure. And from the 1980s onwards, it was feared that nuclear war would cause so much conflagration that Earth's atmosphere would be dimmed by soot. Nuclear winter, as the phenomenon is sometimes called, would much resemble asteroid winter* and has been forecast to cause years of famine. An analysis published in 2022 found that a nuclear war between India and Pakistan could, in this manner, claim more than two billion lives,†[41] while a nuclear war between the United States and Russia could claim five billion.[42] As the Russian aphorism goes: 'The living will envy the dead.'‡

These ghastly prognostications leave us grasping for solutions, for the ways in which we might avert nuclear armageddon. From the early days of the Cold War onwards, a relatively common temptation in the West has been to call on one's government to unilaterally disarm. This policy has an attractive moral clarity. It is true that, if everyone adopted it, the world would be safer. But the advocates of this idealistic view mistake the world they live in for the world they would like to live in. To unilaterally disarm is to invite unilateral harm – inflicted by your adversaries on you and on your allies. North Korea and other rogue states understand this dynamic well and might owe to it their continued existence. We might also owe to this dynamic the failure of the Cold War to become a hot war, the kind of conventional conflict that, if it had been much like the First and Second World Wars, would have cost hundreds of millions of lives.

At its core, the dynamic is a form of the prisoner's dilemma. If every prisoner cooperates with each other – if everyone gives up their nuclear weapons – then they all benefit. But if an individual prisoner breaks ranks, betraying the rest of the group, then everyone else is rendered helpless. So nobody cooperates and we

* See chapter 1.

† India and Pakistan, regional adversaries, developed nuclear weapons from the 1970s and 1990s respectively.

‡ This line has been attributed to Nikita Krushchev, who was premier of the Soviet Union from 1958 to 1964, but no record of it has been found in his speeches or writings.[43]

keep our nukes.* Many billions of people live in countries that are not nuclear-armed, but would nevertheless be direly affected by nuclear war. They remain trapped in a game that they never agreed to play and whose outcome they cannot affect.

Another temptation takes the form of missile defence systems. Israel's Iron Dome, along with the country's other air defence systems, show that relatively short-range missiles can be shot down with a high rate of success. With that in mind, can we reliably shoot ICBMs out of the sky? It would not violate the rules of physics, but it would certainly strain the current limits of engineering. Indeed, the challenge of shooting down a hypersonic missile has been compared to 'shooting a bullet with a bullet'.[45] The US has a missile defence programme dedicated to that mission, yet the programme has intercepted test missiles at a rate little better than that of a coin flip.[46] And hostile missiles are designed to be hard to track; newer versions can duck and dive, and they release decoy warheads. For the time being, missile defences will be easily overwhelmed by a determined adversary, not least because they are so few in number. At the time of writing, the US had 44 ground-based interceptors and 38 warships equipped with missile defence systems.[47] Perhaps, in an unlikely scenario of flawless performance, that would be enough to see off an attack from North Korea. But Russia, according to a recent estimate, has around 1,710 deployed warheads.[48] China is gaining ground. And a single failure to shoot down an incoming ICBM could result in the deaths of millions. British and Continental readers should take little reassurance from the existence of a NATO missile defence programme in Europe; the programme is even less developed than its American counterpart, and Russia exerts continual pressure on NATO to keep it that way.[49] In March 2025, Donald Trump announced that his administration would upgrade the

* Kristian Rönn gives a useful analysis of this situation in *The Darwinian Trap: The Hidden Evolutionary Forces That Explain Our World* (2024).[44]

US missile defence programme, aiming for an Israeli-inspired 'Golden Dome'. As a member of his staff said: 'This is … the monster systems engineering problem. This is the monster integration problem.'

It is possible that laser-based missile defences will, one day, provide a cheaper and more capable alternative to existing missile defences. In 2024, the UK's Ministry of Defence revealed that British scientists, working by night at a military base in the Hebrides, had hit an aerial target with a ground-based laser, DragonFire. The cost of operating the laser, the government explained, 'is typically less than £10 per shot'.[50] But DragonFire is designed to deal with short-range missiles rather than ICBMs, which fly much higher and faster. To bring down an ICBM with a laser, you would have to keep the laser trained on the same spot for several seconds, during which the missile would have travelled tens of kilometres. It would take many years of startling progress for laser-based missile defence systems to non-trivially reduce our vulnerability. And even if that startling progress were to occur, it would not affect the underlying dynamic: that by threatening the world with terrible danger, powerful countries can maintain for themselves a measure of fragile safety. Technological innovation created that dynamic, and that dynamic now sustains technological innovation. For all the tens of billions poured into missile defence, nuclear war remains intolerably plausible.*

* The precise likelihood of a nuclear war remains difficult to judge. In a report issued by the Forecasting Research Institute, nuclear weapons experts assigned a median 4.5 per cent probability to the prospect of a nuclear catastrophe by 2045. (Nuclear catastrophe was defined as an event where nuclear weapons cause the death of at least 10 million people.) Experienced forecasters, whose expertise is in making data-driven predictions about a wide range of topics, put the probability at 1 per cent. Whether 4.5 per cent, 1 per cent or any other plausible estimate, the risk is far greater than should be remotely acceptable. One would hardly board a plane that had a 1 per cent likelihood of falling out of the sky.

One way of reducing the size of the peril is by reducing the size of the stockpiles. I just referred, above, to Russia possessing around 1,700 warheads. The US has about the same. That's easily enough to destroy each other's major cities, but the stockpiles are many times smaller than they were. The reduction occurred partly because of the collapse of the Soviet Union, an event that sparked the race to track down its unaccounted-for nuclear material. (That is the story I told in the introduction: Andrew Weber's hunt for weapons-grade uranium in Kazakhstan.) Principally, though, the reduction was the result of painstaking negotiation. On 31 July 1991, George H. W. Bush and Mikhail Gorbachev sat at a gilded table in Moscow to jointly sign the Strategic Arms Reduction Treaty, known for short as START. It was not the first such treaty between the US and the USSR, but it was an effective one.* Both sides agreed to slash their stockpiles and to permit external verification of their arsenals' size and status. Later that year, the two presidents announced further reductions, via reciprocal unilateral pledges to reduce the size of their arsenals. These pledges resulted in the largest-ever reductions of nuclear stockpiles. For 30 years, the US had had fully fuelled nuclear bombers on continuous alert, with crews ready to leap aboard. But those days were now over.

Treaties do not eliminate the threat of nuclear war, and they are liable to be torn up or ignored. They depend on goodwill and on fair geopolitical winds. Nevertheless, they have the great virtue of being compatible with the prisoner's dilemma dynamic. Both the US and the USSR gained more from signing the treaty than they lost.

But treaties expire. Circumstances change. In 2002, presidents George W. Bush and Vladimir Putin signed a simpler and additional informal agreement that would last until 2012 and did

* An honourable mention to the Intermediate-Range Nuclear Forces Treaty of 1987, in which the US and USSR agreed to eliminate their stocks of intermediate-range and shorter-range missiles that could be fired from land.

not involve on-site verification that both sides were keeping their word. This agreement resulted in a further reduction in warhead numbers. But START was the basis of the mutual reduction in nuclear arms, carrying with it the obligation for each side to permit verification, and it was due to expire on 5 December 2009. At the beginning of that year, the new American president, Barack Obama, agreed with his Russian counterpart, Dmitry Medvedev, that a new treaty was necessary. Such treaties are the product of vast diplomatic and technical labour on both sides. But they are negotiated at a table, face-to-face. Much rests on the individuals charged with leading the negotiations.

That individual, on the American side, was Rose Gottemoeller. The little girl from Ohio, the one who had felt queasy about the Cuban Missile Crisis, was now a diplomat. She was the first woman to lead such a negotiation, but she had a long history in nuclear weapons control. In 1977, the 'Red Telephone' hotline, put in place after the Cuban Missile Crisis, was a landline, with the result that Finnish farmers would accidentally sever it with their ploughs. Gottemoeller, a 24-year-old who could speak Russian, was part of the back-up plan. She had got a job at Fort Detrick, Maryland, where she was part of the team that enabled the two leaders to communicate, if needed, via satellite.[51] Gottemoeller then worked in nuclear policy, first as a US government official, a career in which she assisted with the START negotiations, and, later, as the director of a Moscow-based nuclear policy think tank.[52] When Hillary Clinton interviewed her for the role of chief negotiator for the new treaty, Gottemoeller could hardly have been much better qualified.

But treaty negotiations are not easy, particularly when time is short and the other side is Russian. There is spying, drinking and skulduggery. Much of it overlaps. Early in the negotiation process, Gottemoeller had told the White House that she planned to organise a 'nose touch' – a low-key meeting in which she and her opposite number, Anatoly Antonov, would ensure their teams had come away from preliminary talks with the same sense of what was on the agenda. When Gottemoeller called Antonov to propose they meet, he told her how much he liked the English phrase 'to touch noses'.[53] Gottemoeller saw this as a power play, a deliberate demonstration of Russia's ability to listen in on her phone calls. Unruffled, she continued to use a non-secure line on occasions when she wanted to emphasise certain points to eavesdropping Russians.

Different negotiators have different styles. Male negotiators, writes Gottemoeller in her account of the negotiations, range from histrionic to more measured. Gottemoeller explains that she is generally a very calm negotiator – though sometimes one must demonstrate one's temper. Even at an early point in the negotiation, she had already had to suffer some patronising Russian remarks, delivered at an evening drinks function, about the role that women play as the nurturers of men. Back at the negotiating table, the Russians were defying an agreement between Obama and Medvedev by trying to bring European missile defences, long regarded by the Russians as a threat to their deterrent, back into the discussion. 'This negotiation is going to be about strategic *offensive* forces, not missile defence!', Gottemoeller shouted, pounding the table.[54] The tantrum, she wrote, had the desired effect. 'I didn't need to throw another tantrum for the rest of the negotiations. It was sufficient that I had proven I could do so if I had to.'

What followed were six months of negotiations so stressful that delegates lost teeth and hair. Working brutally hard in Geneva, they contracted high blood pressure, shingles and other stress-related maladies. One official, for a terrifying few moments, stopped talking, hearing and seeing. Gottemoeller was getting three hours of sleep a night, spending the small hours staring at the ceiling and thinking about the treaty and how best to deal with each delegation – both the Russians' and her own. The 'mothership', as the American diplomats referred to the White House, was a demanding taskmaster. On learning that Gottemoeller and Antonov had discussed the idea of a joint statement, external to the treaty negotiation, on missile defences, Obama was 'extremely angry'. Gottemoeller thought she was 'truly done for' as the chief negotiator.[55]

By December, most of the treaty was in place.* What remained was the drawing-up of the treaty's annexes, a laborious process in which officials and translators draw up reams of fine print, including the technical details of weapons inspection procedures, and ensure every sentence is accurate in both languages. But the White House, fearing that the forthcoming midterm elections would make the treaty harder to get through the Senate, was pushing for a rapid conclusion. By this point, both delegations were exhausted. To the horror of the Americans, the Russians announced that they were going to Moscow for a rest period of unspecified length. Gottemoeller, trying to avoid the wrath of the White House, tried to arrange the emergency transplantation of her entire delegation, emergency diplomatic visas and all, from Geneva to Moscow. She told her team to pack her bags – before receiving a phone call from Neil Couch, a colleague at the State Department. 'Rose,' said Couch, 'the Russians were on their way to the airport when they got word that the airport is closed.'[56] The Eyjafjallajökull volcano, which

* For the sake of brevity, I am eliding much of the detail. But I cannot do so without reproducing Gottemoeller's anecdote, related in her book, about 'the constant Russian desire to hang onto equipment, putatively to convert it to other uses. The Russians, for example, said that they wanted to hold onto the "transporter-erector-launchers" that were used to move mobile missiles around and raise them to an upright launch position. They said that they wanted to convert them into construction cranes. As Gottemoeller has it: 'Mike (Elliott, an American official) pointed out regularly that this constructed crane would be no substitute for a purpose-built construction crane; no construction boss in his right mind would want to use one of them. He teased them when they insisted that they couldn't possibly leave the covers off of silos for sixty days so that our satellites would see the process of eliminating them. They complained that cows might fall into them. In that case, Mike said, you could use your converted cranes to haul the cows out of the silos.'

you will recall from chapter 3, had erupted, grounding European air traffic. Against the odds, one source of catastrophic risk had helped address another. Work resumed in Geneva on Monday. The following April, Obama and Medvedev signed the treaty in Prague; it passed the Senate ratification process the following February. The New START treaty, as it is known, restricts the United States and Russia to 1,550 deployed strategic warheads each. On the one hand, the treaty forced the US to dismantle more warheads than did Russia. On the other, it put no constraint on missile defence programmes.[57]

One might be inclined, after the successful completion of such a negotiation, to take a sinecure of some kind. Instead, Gottemoeller became the US State Department's most senior advisor on arms control, before taking a gentler role as the deputy secretary-general of NATO. These days, she teaches at Stanford. I shudder to imagine what it must be like for a student to enter the office of this formidable negotiator without having prepared for her class. 'This negotiation is going to be about your failure to complete your assignment, *not* the death of your grandmother!'

In September 2024, I visited Gottemoeller at Stanford. It was baking-hot on the vast, sandstone campus, another day of endless Californian sunshine; a world away from the snow-storms of Geneva and Moscow. But the thaw in Russo-American relations, the development that had enabled the negotiation of the New START treaty, was long gone. Russia was two years into its war in Ukraine, and its ties with the US were in tatters.

I met Gottemoeller in her office. She was friendly but formidable, exuding a quiet authority. It soon became apparent that I might not have been the only party interested in what Gottemoeller had to say that afternoon. A few minutes into our conversation, she shut the door. 'The walls are paper-thin,' she said, 'even though the outer walls are built like a battleship.' She seemed to be concerned about disturbing her colleagues rather than being eavesdropped on by intelligence agencies, but I was nevertheless curious: was she still of interest to the Russian authorities? Even in the current diplomatic froideur, there remain non-governmental communications between groups in Russian and American security circles; Gottemoeller takes part in these dialogues. Were Gottemoeller's emails being monitored? Her phone?

'Sure,' she said, matter-of-factly. 'Absolutely. I don't like to be listened to, but I don't have any particular compunction about using my knowledge of that to send messages to foreign counterparts, including deterrence messages.'* As an example of the kind of message she liked to communicate in this manner, Gottemoeller referred to the 'idiotic system' that is Russia's 'Skyfall' missile project – an endeavour to build a hypersonic thermonuclear warhead that's propelled by a nuclear-powered engine. 'It's killed developers and engineers rather than foreign enemies. It's such a dangerous system. I've said freely, and often, that they're pouring money down a rathole by spending money on it.'

The advantage of nuclear-powered missiles is that their range is virtually unlimited. They can strike from unexpected directions via convoluted routes, avoiding known radar networks and missile defence networks. But they are risky to build and risky to deploy; airborne nuclear propulsion is less well understood than conventional engines, making for a higher likelihood of crashes and unexpected detonations. The US looked into this technology from 1957, but abandoned the idea in 1964, two years after the Cuban Missile Crisis.[58] All in all, Russia's Skyfall project is symbolic of a kind of belligerence that would best be left in the original Cold War period rather than rejuvenated for the modern era.

It is a project that makes the period of the New START negotiation seem even more fleeting. 'Completely ephemeral,' as

* Journalists live in constant terror of their dictaphones failing. It was a relief to learn that, if the worst happened, I could ask the GRU for a back-up.

Gottemoeller put it. Hard though it might be to imagine, Vladimir Putin had temporarily ceded the presidency, on constitutional grounds, to Dmitry Medvedev. Though Putin was regarded as remaining in charge, Medvedev seemed, at one point in the negotiation, to stand up to his apparent puppet-master in order to get the treaty done. But the moment has passed; Putin has not only taken back the presidency, but amended Russia's constitution so that he can stay in post even longer.[59] His long-persecuted chief critic, Alexei Navalny, is dead; eastern Ukraine is a morass of tank tracks and rubble; an invigorated coalition of non-Western powers has refused to make Russia a pariah; and Medvedev is put in front of cameras to threaten the West.*

This bulldozing of rules and norms has flattened New START, too. As it is, the treaty is due to expire in February 2026. But, in February 2023, Putin announced that Russia would suspend the treaty until the US stopped aiding Ukraine, also demanding that the negotiations encompass the UK and France. 'It's unprecedented,' Gottemoeller told me. 'There is a long, 50-plus-year history of negotiating about nuclear weapons, and Vladimir Putin has interrupted it because of another concern – concern about NATO, support for Ukraine. It's exogenous to the existential threat that nuclear weapons pose to our countries, and to global survival, but that's not the way that Vladimir Putin sees it. So we're in a different time, a different game right now.' The insertion of AI into the decision-making process will make the game even more complex and unpredictable. AI, as I will explain, has begun to knot itself into almost every problem humanity faces.

Given the scale of the threat posed by nuclear weapons, to be a chief negotiator of a treaty like New START must be a

* Britain attracts especial vituperation. In September 2024, Medvedev said that a nuclear response would be 'an extremely difficult decision', but also remarked that 'the pompous Anglo-Saxon imbeciles do not want to admit one thing: any patience comes to an end'.[60]

near-overwhelming task. Gottemoeller, like every diplomat I've interviewed, seamlessly redirected conversation away from herself and toward the bigger picture: in this case, Obama's ambitions regarding arms control. In the back of her mind during the negotiations, she recalled, was the notion that there was a 'benefit to humankind as a whole', but 'front and centre' was the fact that she was serving the president and the American people. It is difficult to imagine Gottemoeller being remotely indiscreet. She did, though, recall a conversation she'd had with Antonov, her opposite number, about the age-old diplomatic phenomenon whereby success has many fathers and mothers, but failure is blamed on the negotiators. 'You get taken out and shot,' as Antonov put it wryly.

Antonov, until recently Russia's ambassador to the US, has now been recalled to Moscow.[61] When she learned of his move, Gottemoeller texted him good wishes. 'We have a pleasant enough relationship,' she told me. At the time we spoke, Antonov's replacement had not been appointed, but Gottemoeller had heard that he or she might be a lower-ranked diplomat. This would be a way for the Russians 'to signal, again, their displeasure with Washington'. It might also be a consequence of the shift in Russia's gaze from west to east. For decades, Russia's principal diplomatic concern was with Washington and the rest of NATO. But today, Gottemoeller told me, Russia wants to develop its relationships with China, India and North Korea. Of those three, at least China will probably be involved in future nuclear arms negotiations between Russia and the US. 'People worry about the "three-player game",' Gottemoeller said, 'but I think that, in general, deterrence-balance among Russia, China, the United States could be accomplished. We can argue about how, but nevertheless, the fact of the Chinese build-up alone does not undermine the basic fundamentals of deterrence.'

The Sino-Russian alliance has been characterised by Putin and his Chinese counterpart Xi Jinping as 'friendship without limits', but it has involved at least one limitation that has been of benefit to the world's nuclear safety. In September 2022, Putin's troops were being chased back across the Dnipro River in south-east Ukraine. This was a humiliating retreat for an army that, in March, had been attempting to take Kyiv. The Pentagon had previously estimated that the chance of Russia using tactical nuclear weapons in Ukraine was between five and ten per cent – until new intelligence caused that estimate to vault to 50 per cent.[62] 'I was very concerned,' Gottemoeller told me. While the White House sought to dissuade Putin from pushing the button, the security establishment held its breath. It did not exhale until Xi Jinping and Narendra Modi, the president of India, publicly warned Putin not to use nuclear weapons. 'They don't ever reprove each other publicly,' said Gottemoeller, 'so this was a huge deal.'

Gottemoeller had never received training in negotiation, but is now comprehensively qualified. It is an art that begins, she told me, with a basic human question: how do I get what I want without using force? Quotidian life, she told me, provides a feedback loop. 'Human beings are negotiating every day of their lives. If you have toddlers at home, you're negotiating how many bedtime stories they're getting. If you have teenagers at home' – Gottemoeller is the mother of two adult sons – 'you're negotiating when they have to bring the car keys back on Friday night.'

It comes down, she said, to your interest and experience in finding mutual solutions and mutual interest, as well as rapport. Gottemoeller told me that when she was working with Antonov, 'I could come in and say, "Hey, how's your dacha construction going?", or "How's your granddaughter doing?", or whatever it is. That kind of minor human interaction can really pay dividends by producing more mutual confidence.' (Specifically, it makes 'the other guy or gal', as Gottemoeller put it, less likely to betray your confidence or make inaccurate reports to his or her government.)

Gottemoeller and I were getting toward the end of our allotted hour. I wondered whether there was anything else she could tell me about how she got the treaty done. For a moment, I glimpsed the near-intolerable psychological conditions she must have had to endure in 2009.

'Well, I'll tell you another thing for damn sure,' she said. 'You need a lot of patience, both in dealing, in my case, with the Russians, and also in dealing with the mothership, as we called it – the Washington interagency and leadership there. It really takes a thick skin, a very thick skin, to be able to just let pressure roll off your back when you have to, because if you succumb' – she paused for a moment – 'you couldn't function.'

We have a dangerous century to endure. The West, Russia and China are overdue a trilateral arms control treaty. Iran might soon have nuclear weapons and North Korea continues to expand its stockpile. The different motivations of these countries' leaders present different challenges in negotiation. Kim Jong-Un, for instance, probably views his own survival as dependent on his nuclear arsenal. So many psychologies, so many considerations – Gottemoeller's successors will have one of the most difficult jobs in the world.

No matter the quality of the next generation of negotiators, it is hard to imagine a future in which there are no nuclear weapons. The scientists who devised them – the Rutherford mentees, their Cambridge contemporaries, the fleeing Europeans, their imitators in Russia and elsewhere – were brilliant intellects. But they were enacting the will of forces far more powerful than themselves, and those forces are as fundamental a feature of human relations as the laws of physics are to the universe. If a species is clever enough to split the atom, but tribal enough to fight, then it will not be long before it dreams up the nuclear bomb.

Nevertheless, since Hiroshima and Nagasaki, humanity has gone 80 years without using the weapons in war. We owe that eighty-year streak partly to diplomats such as Andy Weber,* and partly to cool-headed heroes such as Petrov and Arkhipov. (Petrov died in poverty in 2017; Arkhipov died, in 1998, of kidney cancer, an illness for which the leak of a nuclear submarine's reactor might ultimately have been responsible.) If we want that 80-year streak to be significantly extended, we must wean ourselves off our reliance on luck and quick thinking. Over the long term, our safety will depend on the kind of pragmatism that works within the constraints of human incentives rather than defying them at the last moment, clearing shots off the goal-line. Rose Gottemoeller and her fellow negotiators act on this principle. So do the many technicians and legislators who, in their various ways, have made nuclear weapons easier to track, less liable to fail and harder to dispatch on a whim. Given the secrecy of nuclear weapons programmes, we will never know whom to hail for these small but significant acts of wisdom.

As we go about our business in unbombed cities, we might reflect on some well-thumbed lines of George Eliot's. 'The growing good of the world,' she wrote in 1871, 'is partly dependent on unhistoric acts; and that things are not so ill with you and me as they might have been, is half owing to the number who lived faithfully a hidden life, and rest in unvisited tombs.'

We might also reflect on a story told by a former nuclear weapons launch officer in 2004.[63] He had served in the 1970s. After the shock of the Cuban Missile Crisis, the US government body in charge of ICBMs had been told to install secret codes on the missiles so that they could not be launched without authorisation. Though the US military has denied it, the story

* A 'buccaneer', in Gottemoeller's words; she recalled visiting a former bioweapons plant in Kazakhstan with him in -40°C (-40°F). 'It was evident from his relationship with the Kazakhs who were taking us around, to show us what had been done to shut the plant down, that there was a great deal of mutual respect there. And that's how you get things done.'

goes that the government body, the Strategic Air Command, resented this safeguard. They followed the order, if begrudgingly. Each nuclear weapon would be protected by an eight-digit passcode.

The code? 00000000.

CHAPTER 5

FEEDING THE BENIGHTED

As I came through the desert: All was black
In heaven no single star, on earth no track

James Thomson

You might recall, from chapter 1, the geological whodunnit surrounding the fate of the dinosaurs. It was clear from the fossil record that a mass extinction had occurred, but it was not until 1980 that academic consensus identified the Chicxulub asteroid as the culprit.[1] And in the years that followed, scientists identified the immediate beneficiary of this planet-shaking cataclysm. In the long term, of course, it was mammals who succeeded the dinosaurs as the masters of Earth; in the short term, though, it was fungi.

In 2004, an apparently dry paper, 'Fungal Proliferation at the Cretaceous-Tertiary Boundary', painted an eerie picture.[2] Scientists examining a coal seam in New Zealand had found ancient geochemical traces of conifers, ferns and flowering plants. These traces could be dated right up to the end of the Cretaceous period. This was the moment, in geological terms, at which the dinosaurs were wiped out. The following geochemical layer, the scientists wrote, 'yielded only fungal spores and fragments of nonseptate hyphae belonging to previously undescribed species of fungi'.

Hyphae are the gossamer-like threads through which fungi explore the world around them and absorb its nutrients. ('Nonseptate' refers to hyphae's type of cell structure.) Leave a strawberry out on a countertop and eventually you will see it bound by these delicate white threads. As the threads leach enzymes into it, the strawberry softens and darkens. It is being digested before your eyes, and its nutrients are being shuttled back down the hyphae to feed the fungus' reproductive structures. Soon, the fungus will puff out some spores, seeding another version of itself in a new chunk of organic matter.

Now imagine a world rendered cold and dark by an asteroid impact. The asteroid pulverised so much rock that dust has filled the skies, dimming the light of the sun and killing the plants that feed on it. Underfoot, you can feel the softened remnants of what were once thick ferns. In the gloom, you see rotting tree trunks. The bark of each tree is laced by the hyphae that are slowly but inexorably digesting it. Of the large animals that once stalked this forest, only corpses remain. This dank, silent planet is ruled by mushrooms.

Such is the world that was left by the Chicxulub asteroid. Thanks to the scientists who examined that coal seam in New Zealand, humanity has had a glimpse of that baleful period in the planet's history. It was not just a glimpse of the past, though. As a 32-year-old engineer realised on reading the paper in 2011, it might also be a glimpse of our future. Even if we don't need to worry about asteroids anymore, we are not yet capable, as I have outlined, of preventing super-eruptions. And super-eruptions, of course, are currently far less likely to befall us than is a nuclear war. In each of these scenarios, a multi-year global winter could follow,* causing catastrophic damage to global food chains and worldwide agriculture. As a result, such a winter could kill even more people than the event that caused it.

* As per the previous chapter, there remains uncertainty about the likelihood and severity of a nuclear winter, even in the event of all-out war.

The engineer reading the paper was called David Denkenberger. He had just finished a PhD and was undertaking technical and policy work for a company that makes electronics more efficient. (In this line of work, he contributed to energy-efficiency regulation that saved the US a reported $10bn over seven years.[3]) Denkenberger had been the kind of child who questions everything – 'Why? Why? Why?' – and the kind of teenager who dreams up inventions. As an adult, he had come across the work of Ray Kurzweil, the inventor-turned-futurist. Kurzweil's rapturous forecasts of the union of man and machine come with a flipside: we might not make it that far. Reading Kurzweil's book of 1999, *The Age of Spiritual Machines*, Denkenberger was perturbed by the following lines. 'My own view,' wrote Kurzweil, 'is that a planet approaching its pivotal century of computational growth – as the Earth is today – has a better than even chance of making it through. But then I have always been accused of being an optimist.'

Denkenberger was also reading the early work of Nick Bostrom, a visionary Swede who, in Oxford, was helping to lay the foundations of the study of existential risks. (Bostrom's exploits deserve, and will shortly receive, a chapter of their own – read on.) Like Kurzweil, Bostrom saw the 21st century as a perilous one. He was soon to note – with Denkenberger reading attentively – that there was more scientific literature on dung beetles than there was on existential risk.[4]

Denkenberger, reading these troubling forecasts and wondering what his engineer's skillset could do about them, began to spend time on the Effective Altruism forum. This was a hub where quantitatively-minded charity geeks discussed and evaluated potentially high-impact use of altruistically-intended time and money. The effective altruists were generally drawn from leading English-speaking universities and, at first, in the late Noughties and early 2010s, were concerned principally by poverty in the developing world. (It was in this period, when I was studying at Oxford, that I first encountered effective altruism. I subsequently interned for one of its charities.) By the mid-2010s, the effective altruists – inspired partly by Bostrom's forecasts of rapid technological development – had come to consider the existential vulnerability of humanity to be the most pressing of the world's 'cause areas'. Denkenberger was intrigued. At the same time, he turned to what little scientific literature there was that related to worst-case scenarios for humanity. Thus he found the paper about the fungi that took over the world.

The obvious conclusion from the paper was that we humans, in the event of a global catastrophe, will ourselves be succeeded by mushrooms and their kind. 'And I thought,' Denkenberger told me, 'why don't we eat the mushrooms and not go extinct?'

So began a project that might, if we are first very unlucky and then lucky once more, make Denkenberger and his colleagues the saviours of a battered and reduced humanity. When our planet is next struck by a calamitous earth-dimming event – an abrupt sunlight-reduction scenario, or ASRS, to use Denkenberger's term – the survivors will find their access to food much disrupted. A nuclear war is the most likely culprit here. Perhaps crops will fail; perhaps infrastructure will have been destroyed. There is also the possibility of other disasters: abrupt acceleration in climate change, which would have similarly catastrophic effects on crop yields; a pandemic with an unprecedentedly high death rate; a solar storm that destroys our electronics and power grids; the detonation of several EMPs; or some peril that remains unidentified. (This might sound unlikely, but none of the other potential disasters I've listed were known to humanity until fairly recently.) It's hard to say exactly what these scenarios will look like, or to predict their likelihood with much certainty. But any of them would be nightmarish.

'I thought that the overall field was very neglected,' said Denkenberger. He referred to the Bostrom comment to the effect that, in the early 2010s, there was more scientific literature on dung beetles than there was on catastrophic risks.* 'But then even more neglected were the solutions to the problems. And I thought: *Well, I've an engineering background, and engineers are all about solving problems.*'

Denkenberger is a likeable, industrious sort of character. In his spare time, he attempts to master academic disciplines far removed from his own: law, medicine and other major areas of study. (To save time, he reads while pumping away on an exercise bike.) In his copious back-history of comments on the Effective Altruism forum, Denkenberger discusses topics such the intricacies of nuclear winter, the best ways of staying altruistically motivated, and other aspects of agriculture, science and existential risk – always with faultless geniality. He aims, each year, to do so much hiking that he reaches a cumulative elevation greater than the distance between sea level and the peak of Mount Everest. With these extra-curriculars, he juggles his academic work and his main side-project: trying to ensure humanity can feed itself in the case of disaster.

That side-project began with a partnership Denkenberger struck up with Joshua M. Pearce, a former engineering colleague at Pennsylvania State University. Denkenberger and Pearce set out to create a blueprint via which humanity could, in the event of an ASRS or similarly disruptive occurrence, feed itself. Stockpiling, they realised, was not a viable approach. It would take tens of trillions of dollars to store enough food to last the world five years, which is a plausible length for one of these winters. Simply to initiate that process would cause a surge in global food prices. That surge in food prices, they calculated, would cause starvation in the developing world, thus defeating the point of stockpiling.

* Denkenberger agrees with Bostrom that the direst threat to humanity is super-intelligent AI. But he says the best contribution he can make to human security is via working on resilient foods. And there are, of course, avenues via which AI could make an ASRS more likely – for instance, by accidentally starting a nuclear war.

Humanity would have to cook up its own menu. Swapping their lab coats for metaphorical chef's whites, Denkenberger and Pearce got to work. In 2014, they published a recipe book of sorts – an almanac that explained how a benighted species could feed itself. It was called *Feeding Everyone No Matter What: Managing Food Security After Global Catastrophe*, and Denkenberger and Pearce dedicated the book to their children. 'May they never need to use it,' they wrote.

Before I lift the cloche, you should know that the premise of the book is not wholly pessimistic. 'If you ingest a lot of mainstream media,' wrote Denkenberger and Pearce, 'you are forgiven for wondering at this point "What about bees?".' Journalism of the 2010s was full of 'sensationalized bee death stories', they wrote, but humanity need not fear. The loss of all pollinators would cost humanity only three to eight per cent of our agriculture, they wrote, referring to a study from 2009.[5] It's also worth noting that not all pollinators are bees. 'The complete loss of bees, although extremely unfortunate, would not be an agricultural disaster of the magnitude focused on in this book.' Ditto the degradation of topsoil, and other worries. 'Humanity should be more than nimble enough to stop all of these problems from becoming global catastrophes.'

On that relatively uplifting note, on to the menu. In the shorter term, we can grind and press leaves before coagulating the resultant liquid into something vaguely edible. (Denkenberger has tried leaf protein concentrate. 'It kind of tastes like spinach. It's very green.')

Hungry, hungry humans might be able to do the same with inner bark or ground sapwood, which is the wood located closest to the bark. We could use ruminants – cows, sheep, goats and so on – to turn leaves, grass and other plant matter into milk and meat. It takes years, though, to multiply the size of a flock. Ruminants alone will not be able to feed the planet.

Mushrooms, naturally, play a role here. Viewers of the recent post-apocalyptic TV series *The Last of Us* will recall buildings colonised by a monstrous fungus that preys on the remaining humans. In the Denkenberger and Pearce scenarios, we encounter a happy inversion of this scenario. The scientists imagine us repurposing offices and industrial buildings as nurseries for white button mushrooms, exploiting a eukaryote species for our benefit. 'The mushroom solution can quickly ramp up to provide all of human food needs,' wrote the authors, albeit with the acknowledgement that this would be a 'fairly monotonous diet'. We could also grind up some wood and introduce to it some bacteria that turn it into a sugar – thus creating a supply of sugary water that could, in theory, make for a somewhat pleasant drink. 'Following this model,' wrote Denkenberger and Pearce, 'at least some of the humans would continue to survive off of fairly attractive foodstuffs': steak and the fizzy pop one could concoct by carbonating the sugary water.

And we would probably be able to surf and turf. Marine food chains would be less badly affected by an ASRS than those on land. We would still be able to fish, and because seaweed can survive off atypically dim sunlight, we might find ourselves ramping up our production of it. As the authors point out, though, seaweed is prohibitively high in its water content, which makes it inefficient to transport. We might have better luck converting fossil fuels into edible oil or into sugar; turning methane into single-cell protein; or farming rats that will eat biomass that has been partly broken down by bacteria and fungi. A more luxurious alternative to the rats would be chickens.

Pursuing this line of research, Denkenberger told me, has not resulted in his experiencing any dark dinner times of the soul. 'Even though I work on these big, bad scenarios,' he said, 'I really focus on what we can do about them. And I think, overall, it's a hopeful vision to say: "Well, if we cooperated, we really could feed everyone in these terrible scenarios."'

The book was of particular interest to an English humanitarian worker called Ray Taylor. Taylor is a Quaker, the Quakers being a Christian movement known for its storied history of pacifism and charity. His CV included his part in the operation that sought to feed Madagascans after the country was struck by a violent cyclone in 1999. Taylor's career, in other words, had left him needing no convincing of the fragility of the world's food supply. He agreed that the whole planet should have a plan to feed itself in the event of a catastrophe and believed that the book offered a good blueprint. He found Denkenberger via Facebook and the two men spoke on the phone. Denkenberger struck Taylor as upbeat, intelligent, slightly shy – and possessed of a similarly deeply held ambition to do good.

The blueprint would be no good, they agreed, if it resulted in no action. Left to their own devices, national governments probably wouldn't be interested in preparing for events unlikely to take place within one or two election cycles. As a result, the need for preparedness was likely to go unmet. All this amounts to what's known in the trade as a 'coordination problem' – where coordination makes sense, but is difficult to achieve. Some coordination problems, such as the need to address climate change, require so much change to the machinery of society that they are fiendishly hard to solve. (Later, I will relate the story of a piratical attempt to unilaterally solve the climate crisis.) Other coordination problems, though, seem more plausibly fixable, at least in the eyes of individuals of a particularly buccaneering spirit. Denkenberger and Taylor were far-sighted enough to agree that an organisation should start making the ideas of the book more practicable – and buccaneering enough to found it themselves. Like other members of the anti-catastrophe league, Denkenberger and Taylor had sufficiently adventurous temperaments to do something about a problem that few others saw. And where Denkenberger had the engineer's eye for a practical solution, Taylor had an intricate understanding of the webs of government agencies and non-governmental organisations with which this new body would need to work. 'My role,' Taylor told me, 'was, to some degree, getting David out of his academic and writing environment [and] into this professional world.'

They called their new organisation the Alliance to Feed the Earth in Disasters, known for short as ALLFED. (That ingenious coinage was Taylor's.) Its goals were to identify, pilot and prepare resilient food alternatives, and to help governments and communities implement those solutions. Denkenberger's Dank Burgers was open for business.

Denkenberger was so convinced of the importance of ALLFED's work that he donated 50 per cent of his academic salary to the new project. 'My philosophy,' he wrote in an Effective Altruism forum discussion about making regular charitable donations, 'is to earn like upper middle class, live like middle class, and donate like upper class. One can typically accomplish this by roughly maintaining the consumption per person that one has earlier in life (e.g. college or grad school). Sure, there is temptation to have consumption creep as is happening in most of one's peers, but it is not technically difficult like rock-climbing, or nearly as bad as living with hunger on a diet.'[6]

His friends and family did not have the same kind of zeal, but nor did they consider Denkenberger's work to be ghoulish. Instead, he reports, they are generally supportive. 'Well, I'm glad someone's actually working on this,' people tell him.

ALLFED took on engineers, economists, scientists and more. (The team works remotely and is scattered across every continent except Antarctica. Its preponderance of NATO citizens, however, has prompted internal discussions of the merits of scattering even more widely.) It is a relatively small team, but is known for its prolific research output. ALLFED has evaluated, in more detail than Denkenberger and Pearce were able to muster in their book, how we could turn various industrial chemicals, at scale, into edible material. It has re-evaluated the potential of seaweed, finding that its production could be increased massively.[7] (Enjoy your post-apocalyptic sushi rolls.) It has advised governments, offering national plans that take into account local agriculture and industry. ALLFED has worked particularly closely with the Argentinian government* and has also undertaken research on behalf of NASA. Highly efficient food, it turns out, is useful for astronauts as well as the benighted.

Some of the most arresting work has been on cellulose: how we could turn all that dead plant matter into sugar. Paper mills, ALLFED staff realised, are already well-suited for processing wood – and could be rapidly repurposed to turn that wood into food.[8] The team hopes that, over the next few years, it will be able to pilot this work using a real paper mill.

Given all this talk of resilient food, as it's called, you might well be wondering what the staff eat when they meet up. When I attended an ALLFED-hosted dinner in 2022, we ate Mexican food, but when eating as a team, the staff has experimented with more on-brand dishes such as miso soup. Denkenberger himself, having looked into the post-apocalyptic viability of subsisting on insects and arachnids, has put his mouth where his money is. 'Though I have eaten a scorpion,' he told me, 'it is not a promising resilient food – favourable ones eat fibre.'

At the dinner in 2022, I met Mike Hinge, the organisation's senior economist. When I interviewed Hinge the following year, I asked him what it's like to sit down at one's desk each day, sip one's coffee and envisage the human race being struck by something truly horrendous. The sense of being able to do something about the problem, Hinge told me, is uplifting – though he acknowledged that those who work in the field of existential risk sometimes become affected by emotional burnout. 'I feel genuinely privileged to be able to do it,' he said of his work. 'But I think that, to a degree, it can be a bit like altitude sickness,

* The collaboration with Argentina grew out of ALLFED's work with a Spanish-speaking observatory of global catastrophic risks. ALLFED has contributed to Argentina's disaster preparedness policies, and in 2024 ran a training exercise on global catastrophic food failure in conjunction with the Argentinian government.

where you just don't know until you climb the mountain whether or not you're susceptible. It's no judgement on other people if they find this work to be morbid or stressful.'

That billions of people eat multiple nutritious meals every day is an extraordinary feat as it is. The logistics of this process are remarkable, putting Hinge's job somewhere between maddening and fascinating. 'The food system is incredibly complicated in itself,' he says, 'and it connects to everything else such that each time you pull the threads, it goes on forever. You end up encountering the dynamics of each individual crop, the complications of how markets work and the relationships between farmers and the government in each country.'

The problem for ALLFED researchers, said Hinge, is 'narrowing down to a manageable scope. Lack of material is never a problem. It's always a tragedy, and David always hates it, but we have to round down and try to focus on a defensible core.'

There it is again – Denkenberger's relentless urge to ask 'Why? Why? Why?'. If there is a hallmark of his research output, it is the application of this wide-eyed inquisitiveness to problems that others might ignore. When Denkenberger was an energy-efficiency consultant, he worked in Durango, Colorado, near the site of a colossal extinct supervolcano: La Garita Caldera. ('Maybe the only ever VEI9 eruption,' Denkenberger tells me.) *Why shouldn't we try to stop super-eruptions?* wondered Denkenberger. He put this question to a geology professor, Robert W. Blair, at the local university. 'He thought I was crazy at first, thinking that we could do anything about influencing what happens with a supervolcanic eruption, but I eventually won him over.' In 2018, the two men co-authored a compendious analysis of the various methods by which supervolcanic eruptions might be prevented or mollified[9] – this analysis being the one that I referred to in this book's chapter on super-eruptions. (Yes, that was the same David Denkenberger.)

Denkenberger has continued in his academic work, volunteering as ALLFED's part-time director of research while still donating half of his salary. His more recent research bears the same stamp as his older work. You might recall, from this book's chapter on solar storms, a paper whose authors were trying to work out how a household might stay warm in the event of massive disruption to local or global energy infrastructure. Why should people have to shiver? What if there was a way to burn wood indoors without using a fireplace? The report* assessed the viability of converting household appliances to wood-burning stoves, also analysing the scalability of such conversions. 'Be clever or be cold,' warned the report's four authors – of whom the fourth was one David C. Denkenberger.

Given that he has also worked for the Global Catastrophic Risk Institute, Denkenberger's is a life preoccupied by doomsday scenarios. What does his wife make of it? 'She doesn't want to hear about catastrophes all the time,' he says. 'She says she hears enough just from me being on the phone. But I think it's an important thing for future generations, and our kids' generation in particular.'

One wonders, then, what doomsday preparations, if any, Denkenberger has made specifically for his family. In their book, he and Pearce refer to preppers – the communities of modern millenarians whose hobby is to prepare themselves (hence the name) for society-scale disaster. Preppers store food, water, wind-up radios, potassium iodide,† even gold bars that might come in handy for post-apocalyptic bartering. Denkenberger has stored a fortnight's worth of food and water for his family, but not much more.‡ His hope is that the survivors of a dire event will cooperate, thus ensuring that the *Mad Max* scenario remains confined to fiction. Denkenberger is much less an

* A preprint, at the time of writing.

† When you're exposed to radiation, as you might be in the event of nuclear war, potassium iodide lowers your risk of contracting thyroid cancer.

‡ British subjects are instructed by the government to have three days' worth of food and water.[10]

Immortan Joe (the *Mad Max* warlord who maintains brutal hegemony over a desertified continent's water supply) than he is a Norman Borlaug (a gentle agronomist who has been credited with saving a billion people from starving to death).[11]

Borlaug's innovations, such as his creation of high-yield forms of wheat, resulted in huge increases in agricultural production at a time when the world's population was skyrocketing.* He is a hero to the ALLFED staff, who recognise each other's above-and-beyond contributions with the award of a medal – a bronze duplicate of the Congressional Gold Medal that Borlaug received in 2007.[12] Denkenberger has been presented with one of these duplicates, which now hangs in his office at the family home in New Zealand.

Scholars of catastrophe will note that New Zealand is widely regarded as the world's best place in which to see out a global disaster. Denkenberger is there for work, but readers considering the purchase of a catastrophe bolthole might be interested to know that New Zealand's virtues, in this respect, are its plentiful supply of fish and sheep, its temperate climate† and its distance from large landmasses, from which trouble might spill over if survivors are fighting for resources.‡ (Denkenberger emphasises that there should be enough food to go round.)

Perhaps the most detailed treatment of this topic comes from Matt Boyd and Nick Wilson, two Kiwi academics.[14] In 2022, I saw Nick Wilson, from the University of Otago, present to a

* Low crop yields were not an existential risk to the whole of humanity, but it would be graceless not to admit Borlaug to the anti-catastrophe league.

† It's helpful, in this respect, to be surrounded by sea.

‡ Several billionaires will be flying to New Zealand in the event of a catastrophe. One of them is Sam Altman, CEO of OpenAI.[13] Altman's record as a corner-cutting AI accelerationist does not suggest unwavering dedication to planetary safety, but those of us exposed to the risk (that is, all of us) can comfort ourselves with the thought of him hunkering down in a cosy bunker.

conference the pair's ranking of the world's safest islands in the event of a global catastrophe. New Zealand, naturally, was top. Wilson told me that the UK would be third in the European area, lagging behind Iceland, which has 'massive marine resources', and Ireland.[15] Disappointingly, the UK would not make the worldwide top 10, largely because of its closeness to the mainland.

Humanity, then, has identified the foodstuffs and islands that best equip it to see out a disaster. What next for the survivors? And, first of all, do they matter? One might imagine that a disaster that eradicates every single one of us (scenario 1) is not that much worse than a disaster that wipes out 99 per cent of us (scenario 2). Even a minuscule remnant of humanity, though, might be able to rebuild. Scenario 2 is therefore much more desirable. The philosopher Derek Parfit, an indirect influence on Denkenberger and many others in the catastrophic risks arena, made such an argument in his book *Reasons and Persons*. (The book was the centrepiece of his project to rescue morality from godless nihilism. I will return to Parfit, who was one of the great philosophical eccentrics, a little later.) 'The Earth will remain inhabitable,' Parfit wrote, 'for at least another billion years. Civilization began only a few thousand years ago. If we do not destroy mankind, these few thousand years may be only a tiny fraction of the whole of civilized human history.'

Our benighted survivors, then, bear two heavy burdens: not only the mouldering logs that they are carrying through the gloom to their local paper mill, but the future of our species and of everything we value. This band of stragglers probably won't have in their number a perfect distribution of engineers, farmers, surgeons and writers of popular non-fiction. They will have to conjure up the know-how that undergirds everything we take for granted in modern society – and they are unlikely to have access to the internet. With this problem in mind, the British writer Lewis Dartnell has compiled, in his book *The Knowledge: How to Rebuild Our World from Scratch*, some instructions for various civilisational essentials: preserving food, designing car engines, generating power and so on. Books, of course, are a fragile substrate for what would, ideally, be ineradicable information – 'I think you want an apocalypse-proof, unhackable Kindle,' Dartnell once remarked* – but projects such as *The Knowledge* make us more resilient.

And we will need to be resilient – and then some. On the one hand, humanity has got out of some pretty hairy situations in the past. Recent analysis of our genome suggests there was a period of 100,000 years in which the total population of humans might have dipped, at times, to as low as 1,000.[16] On the other, surviving these periods, and eventually managing to reproduce our way out of the bottleneck, is no guarantee of regaining our current abilities.

The Antikythera mechanism is a case in point. In 1900, this complex, geared contraption was rescued from an Ancient Greek shipwreck. It demonstrated a greater level of technical sophistication than we had previously ascribed to the Greeks. The mechanism is now believed to be an astronomical calculation machine, housing 30 gear wheels in its shoebox-sized case.[17]

To learn of the Greeks' forgotten prowess has been a joy to classicists. But a more unsettling element of this story is that the Greeks – like other civilisations that followed our species' natural tendency to mimic and predict the heavens' rhythms using mechanical technology† – never turned this prowess into industrial equipment. They had the ingenuity to start an industrial revolution, but they never used it. It is the civilisational equivalent of having logs, kindling and matches, and using this ensemble to play pick-up-sticks.

* This remark was made to Bryan Walsh, my fellow apocalypse scribe, and published in Walsh's book *End Times: A Brief Guide to the End of the World*.

† Derek J. de Solla Price tells this tale with verve in his book *Science Since Babylon*.

Our modern world is imperfect, but we feed and water billions of people, we develop remarkable new technologies, and we are driving poverty into inexistence. This could have happened centuries earlier, but it didn't. Instead, we endured two millennia of famine, war and poverty. There is no deterministic path to industrialisation. By the same token, civilisation might well be a one-time occurrence. Should the flame gutter, the work of Denkenberger, Taylor, Dartnell and others might prevent it from going out entirely.

CHAPTER 6
TEARING OFF THE HAIRSHIRT

With throats unslaked, with black lips baked,
We could nor laugh nor wail;
Through utter drought all dumb we stood!
I bit my arm, I sucked the blood,
And cried, A sail! a sail!

Samuel Taylor Coleridge, 'The Rime of the Ancient Mariner'

I alluded earlier to the phenomenon of – for want of a better term – thick-skinned weirdos. These are the people whose jeremiads call civilisational attention to an impending danger. Among these iconoclasts are the 19th-century gentleman scientists who proved that asteroids were real; the 20th-century physicists who warned, fruitless though those warnings turned out to be, of the peril of the thermonuclear bomb; and, as I will demonstrate, the 21st-century philosophers and computer scientists who drove early work on the development of safe AI.

Not all millenarians are correct. The end is rarely as nigh as is claimed to be. If I had a pound for every time someone incorrectly predicted the end of the world, I'd be rich enough to build an apocalypse-proof bunker spanning the length of New Zealand. Altogether, though, those thick-skinned weirdos – at least those whose arguments come from the scientific method rather than numerological analyses of sacred texts – do us a great service. They endure suspicion, hostility and ridicule in order to bring our attention to a hazard.

If only humanity had listened to Svante Arrhenius, the Swedish physicist and chemist who, in 1896, predicted that the burning of coal would result in global warming of 3 to 4°C (about 5 to 7°F).[1] If only, given a second chance at addressing climate change in good time, humanity had listened to Guy Callendar. In 1938, Callendar explained to a panel of scientists at the Royal Meteorological Society that humanity's use of fossil fuels was already affecting the climate and would go on to affect it further.[2] Callendar was the son of an inventor, one of four boys, and grew up in a chaotic-sounding home in which one of his brothers blew up a greenhouse while trying to make TNT. As a result of an accident involving a pin and the same brother, Callendar had lost his sight in one eye.[3] But it was the Royal Meteorological Society, whose panel dismissed Callendar as a dilettante, that was blind; he had managed to peer into humanity's future. That future – our present – is a world in which the planet's temperature is rising at a rate hitherto unseen in 65 million years.[4] Rightly, there is general global recognition that we ought to arrest this trend – and perhaps, though it might take some environmental dark magic, reverse it.

But first, let there be no denial of the miracle of fossil fuels. They come ready-made; they are fantastically dense in energy; they are wonderfully portable, able to be stored indefinitely; and they require no electronics to be put to work. Fossil fuels remain more accessible sources of heat, light and motion than almost anything we have devised for ourselves. It is difficult to imagine that another miracle, the wrenching of billions of people out of poverty in the 20th century, would have been possible without fossil fuels.[5] It is a stroke of inordinate cosmic luck that our species, appearing several billion years into the existence of life on Earth, found the planet so richly endowed with the energy equivalent of free money.

But perhaps it was more of a loan – one that we are now, one way or another, having to pay back.* Eventually, belatedly, and despite the efforts of oil and gas companies, the argument from Arrhenius, Callendar et al became impossible to ignore. Today, there is no need for me to sermonise about the peril that is climate change. You already know it all. You know that the curse of fossil fuels is that, when burned, they release greenhouse gases such as carbon dioxide. These greenhouse gases lurk in Earth's atmosphere and retain heat that would otherwise escape into space. Need I spell out all the consequences? Hotter summers, warmer winters, violent weather, melting ice caps, rising seas, air pollution, mass extinctions and a Swedish schoolgirl, with tears in her eyes, berating United Nations delegates with the words: 'How dare you? You have stolen my childhood and my dreams with your empty words.'[7]

Greta Thunberg received applause for her speech, a result that was testament to the decades of work put in by her fellow climate advocates. These thick-skinned weirdos, to use the technical term, achieved mainstream cut-through on 23 June 1988. That was the day when the director of NASA's Institute for Space Studies told a US Senate Committee that the 'greenhouse effect', as it was then called – a greenhouse even more dangerous than the one that Callendar's brother had blown up – was measurably changing earth's climate.[8] The following day, the front page of the *New York Times* shared the news with tens of millions. 'Global warming has begun, expert tells Senate' went one headline. 'Sharp cut in burning of fossil fuels is urged' went another.[9] That year, the UN co-created the Intergovernmental Panel on Climate Change (IPCC).

Over the years that followed, huge swathes of people were

* The effective altruist philosopher William MacAskill argues that we should leave as much coal, gas and oil in the ground as we can – in case civilisation collapses.[6] In this dire scenario, the remnants of humanity will be very grateful for easily-accessible energy sources.

inspired to reduce their reliance on fossil fuels and to lend their voice to policies that might compel society to do the same. One of those people was a young German called Johannes Ackva. As a teenager and student in the Noughties, Ackva flew as little as he could and chose not to learn to drive. For related ecological reasons, he also tried to limit his use of water, lest it run out.[10] Many will be familiar with this attitude of environmental parsimony; many know what it is like to wear the coarse fabric of the hairshirt.*

Two decades on, Ackva works for a non-profit called Founders Pledge. Here his research guides the philanthropy of entrepreneurs who want to use their wealth to address the climate crisis. And he remains an environmentalist. He is still vegetarian and he still does not drive, but his hairshirt weighs a little lighter, feels a little less itchy. As he explained to me, two things have changed.

One was that the world's climate change data looks a little less bleak than it used to. 'You have to be more optimistic than you would have been 10 years ago,' Ackva told me, 'and even five years ago.' For many years, the IPCC's scientists had forecast that global temperatures were likely to increase by between 1.5°C (the best case) to 4.5°C (the worst). But, from 2021, the scientists' forecasts had contracted: the lower end of the likely range was a 2.5°C rise in temperature, the best estimate was 3°C, and the higher end of the likely range – the worst-case, in other words – was a 4°C rise.[11]

Even a 2°C rise would be disastrous. More droughts, more heatwaves, higher sea levels, more extinctions.[12] A 4°C rise would be far worse. Many people believe that climate change will eradicate humanity entirely; such was the verdict of 31 per cent of the Americans surveyed by a major pollster in 2022.[13]

* A hairshirt, of course, is a garment worn by self-abasing priests of yore. Hairshirts are rough, stiff and unpleasant to wear. Hence the old idiom: people who deny themselves something are 'hairshirted'.

We should be troubled by the unexpectedly high temperatures of recent years.[14] In 2024, the planet's average temperature was more than 1.5°C higher than it was before the Industrial Revolution – a mark that the world's governments had resolved, in the Paris Agreement of 2015, to try not to exceed.[15] We have failed to meet our targets and we have incurred terrible suffering as a result.

That said, the figures do not suggest that the public's existential worries are realistic. Even if the IPCC projections significantly underestimate the extent of future warming, much of the planet will remain habitable. It will be a tragedy, but one that our species will survive.

Nevertheless, further warming, even to a more modest degree, will cause great suffering. And its effects will reverberate through other domains, too. 'Climate makes other risks more salient and stronger,' says Ackva, 'mostly through political destabilisation. Food insecurity can lead to political instability, leading to political risk and to civil war.'

But there is another reason for qualified optimism. This is the second of the broad changes cited by Ackva, and it is the rate of technological progress. 'We can see from what we've been doing with solar, or with electric cars, that we know how to fundamentally solve this problem.'

The price of solar energy has fallen almost at the speed of light; the cost of solar panels has fallen 90 per cent in the past decade alone.[16] Renewables are helping to push coal and natural gas out of power grids. In 2009, 74 per cent of British electricity came from fossil fuels; in 2023, it was only a third.[17] Over the same period, the share of British electricity coming from renewable sources – solar, wind et al – has gone from two per cent to 40 per cent. Other grids, such as that of California, are breaking clean-energy records of their own.[18]*

* For example: there was a point in the afternoon of 18 April 2024 where solar power met 86.4 per cent of the state's energy demand.[19]

In short, we are living through the most rapid energy transition in history. The transition is imperfect: it is incomplete and it is expensive. It is incomplete because renewables are dependent on the sun shining and the wind blowing, and because we do not have adequate battery technology to store vast amounts of electrical energy for months at a time. It is expensive partly because of all the infrastructure that needs building, be it car-charging facilities, wind farms or the apparatus that connects both of these things and more to our power grids. For the UK's nauseatingly high energy prices, we can also blame a regulatory regime that makes new power plants very nearly the most expensive, megawatt for megawatt, in the world.[20] Well-meaning legislation has resulted in provably safe designs being beset by Nimbys and waylaid by environmental impact assessments that are more than 30 times longer than the complete works of Shakespeare. This regulatory quicksand results in absurdities such as the 'fish disco' of the forthcoming Hinkley Point C plant, an over-elaborate and highly expensive attempt to keep fish out of the reactor's piping by blaring continuous high-pitched noise at them.[21] The number of fish affected is small, but the cumulative costs of these measures is vast.

When bemoaning the cost of British nuclear energy, we must also cite the country's sprint, mandated by law, to reach 'Net Zero' by 2040. Few transformations of any worth ever occur painlessly. Imperfect though it has been, the recent progress in the energy transition is demonstrative of a truth that many environmental activists would prefer to ignore. It is common for such activists to advocate for 'degrowth', whereby we shrink our economy in order to constrain our emissions. In this view, degrowth is to society as carbon footprint reduction is to the individual. The less we consume, this logic goes, the better. Consider, for a moment, the apocryphal story of Kim Kardashian taking a private jet to Paris, the city in which the world agreed to limit global warming to 1.5°C, in order to eat a slice of cheesecake.* You might find yourself nodding in sympathy to the chants of 'Just stop oil!', your skin beginning to itch in yearning for the hairshirt you believe we all deserve.

Resist the temptation. Irrespective of wanton dashes for cheesecake, degrowth must not be the solution. It is a philosophy that fails to take account of two very powerful forces. The first is the ability of economic growth to improve lives worldwide; the second is the inability of individual humans to reliably behave selflessly. As with other anthropogenic† threats to humanity, climate change is a product of concentrated rewards (delicious cheesecake for Kim Kardashian after just a few hours of luxurious flying) and dispersed punishments (a fraction of a fraction of a degree of warming, shared between everyone alive). The most feasible approaches to climate change take account of these forces. They divert the river rather than attempt to dam it.

Meanwhile, degrowth campaigners block roads and splatter priceless artworks with tomato soup. They might consider themselves to be members of the anti-catastrophe league, but they lack the requisite pragmatism. They ignore the existence of those powerful forces and they ignore the ability of technological progress to drive down our emissions without defying those forces. If they had their way, we would be getting poorer and poorer, less and less able to maintain healthcare systems and food supplies. Degrowther organisations are not the kind that Ackva recommends his philanthropists support.

Instead, he recommends that they fund innovation: inventions, and government action, that take the battle beyond wind farms and electric cars. 'We need to make sure,' he told me, 'that we also do it for cement, that we do it for [the manufacture of]

* The truth of the matter, according to the fact-checking website Snopes, is that Kardashian flew from Los Angeles to Paris to eat *several* dishes, of which the cheesecake, from Hotel Costes, was but one.[22]

† Human-generated.

steel, that we do it for long-range transport. But this is not rocket science. This is, in principle, doable.'

It has been doable for longer than we might care to admit. You will recall the power of nuclear fission, in which the splitting of an atom releases enough energy to level a city. Nuclear fission can also be used to fuel power stations. After the Second World War, governments on either side of the Iron Curtain built nuclear power stations by the dozen. 'It is not too much to expect,' said Lewis L. Strauss, chairman of the United States Atomic Energy Commission, 'that our children will enjoy in their homes electrical energy too cheap to meter, will know of great periodic regional famines in the world only as matter of history, will travel effortlessly over the seas and under them and through the air with a minimum of danger and at great speeds, and will experience a lifespan far longer than ours.'

Nuclear power stations are gargantuan structures, complete with puffing chimneys atop blue-glowing reactors, and they are expensive to build. They also produce radioactive waste. This effluent is relatively unproblematic in itself,* but conceptually unattractive to the public.

Even less attractive to the public is the phenomenon of nuclear meltdown. Normally, a nuclear power station's reactor core, which is where fission takes place, is kept cool.[23] This ensures that the reaction is kept under control. But the world does not have an unblemished record in this regard. Chernobyl is an infamous example – the worst meltdown event of all time. The power station was built poorly, and its operators, conducting an experiment on 25 April 1986,[24] turned a mistake into an accelerating series of errors.[25] With the nuclear chain reaction getting out of hand, the technicians drove more than 200 control rods, which absorb neutrons and thus slow a reaction, into the reactor. But the unusual conditions meant that the graphite tips

* It remains radioactive for centuries, but can be safely stored in casks made of thick concrete.

temporarily increased the reactor's reactivity. An explosion blasted tens of tons of radioactive material out of the reactor core. Exposed to the air, the chain reaction continued – an infernal sludge belching out radioactive gas. The town was evacuated, but the meltdown's death toll is estimated to have been 300 to 500.[26] These figures are lower than one might imagine, but the victims of the Chernobyl disaster stick much more indelibly in the mind than do the abstract millions who die of fossil fuel-related air pollution each year.[27] One thinks of the helicopter crews who were asked to put out the reactor fire by dropping sand and clay on it from above. These dutiful men absorbed extraordinary amounts of gamma radiation as they hovered low over the fire. Many of them were dead within weeks, suffering from the same symptoms of radiation sickness as did the survivors of the Hiroshima and Nagasaki bombings.

Despite the Chernobyl disaster, and despite smaller meltdown events at Fukushima* and Three Mile Island,† nuclear power is much safer than fossil fuels. The use of nuclear energy results in 99.8 per cent fewer deaths than does the use of coal, 99.7 per cent fewer than oil and 97.6 per cent fewer than gas.[28] Fossil fuels are responsible for perhaps five million deaths per year,[29] largely as a result of the particulates they emit when burned – tiny particles that, by burrowing their way into delicate tissue, kill us via heart disease, Alzheimer's and other ailments.

But these deaths are not as dramatic, nor as obviously causal, as death via radiation sickness. And so we have throttled nuclear power with regulation. Particularly in the US and the UK,

* In March 2011, an earthquake off the coast of Japan caused a tsunami that hit four nuclear power stations in Fukushima prefecture. Two of them had their cooling systems so badly disrupted that they went into meltdown. Thanks to the selfless work of staff, contractors, military and emergency services, nobody died of radiation poisoning.[33] That said, the wrecked reactors remain highly radioactive, and clean-up could take more than a century.[34]

† A partial meltdown in 1979 that caused no injuries or deaths and has been the subject of much hysteria.[35]

nuclear power stations take decades to build and arrive at exorbitant cost. Forty years after it was first proposed,[30] the UK is due to complete the Hinkley Point C power station, putative home of the fish disco, in 2029.[31] But the energy the facility will produce is forecast to be more than four times more expensive per megawatt, when taking into account the cost of its construction, than that of its cousins in South Korea.[32]

Nuclear power, to reiterate, is not perfect. It is expensive to set up power stations, far more so than it is to burn fossil fuels. Further complication comes from nuclear power's dependence on fissile materials, such as our old friends uranium-235 and plutonium-239. As you might recall, these materials can be used to produce nuclear weapons – and are therefore subject to stringent international regulations. These obstacles notwithstanding, it is depressing to think how much clean and reliable grid power the world has sacrificed by eschewing nuclear energy.* To some degree, climate change has been a choice.

This dismal situation appears to be on the turn. Advocates for nuclear energy are slowly winning the argument. A recent Miss America, holder of an office whose occupants are rarely given to controversial statements, has become a public advocate of nuclear energy.[38]† And Western economies, after years of neglect, are now making long-overdue investment into nuclear power. The enormous nuclear power stations of yore look likely to be replaced by small modular reactors, which can be rustled up at speed, en masse, out of pre-fabricated components. At the time of writing, only China and Russia were harvesting grid power from small modular reactors – of which they have one each.[39]

* One recalls the German Green Party's pigheaded campaign to rid their country of nuclear power. They met this goal in 2023,[36] at which point the war in Ukraine was causing energy bills to skyrocket. As a result, Germany had to fire up several of its old coal-fired plants. The German Greens are warmly invited to the first meeting of the pro-catastrophe league.[37]

† Grace Stanke, winner of Miss America 2023, studied nuclear engineering at the University of Wisconsin.

But more are on their way. The Polish government has approved the construction of 24 small modular reactors,[40] while the South Koreans have authorised a small modular reactor design of their own.[41] In the US, tech companies are commissioning small modular reactors in order to feed the energy-hungry maws that are their data centres. And even in the UK, the global capital of infrastructure-throttling regulation,* several small modular reactors should be online by as early as 2027.[42]

Large-scale investment in this unproven technology is an example of what Ackva calls 'big bets' – calculated risks that, taken together, add up to good odds. 'Right now, collectively, we're making a really big bet on renewables in particular,' he told me. 'We know that renewables are cheap, but we don't know that it's easy to scale the energy system to something that's close to 100 per cent renewable.'

As Ackva argued, it makes sense to hedge those bets. If conventional, lithium-based batteries aren't enough to sustain a power grid that runs on renewable energy, then perhaps we can spur the invention of new forms of power storage. (In 2023, I spoke to some enterprising scientists who had managed to power a lightbulb with energy they had stored in concrete – raising the alluring prospect of houses and roads that store renewable energy.[43]) If renewables still don't work out, and nor do small modular reactors, then perhaps geothermal will. (This tantalising technology will be the subject of the next chapter.) Or perhaps the best bet will be fusion.

Fusion, as you will recall, is what powers the Sun, as well as thermonuclear bombs. Here is the method most often attempted by humans. Subject deuterium and tritium† to extreme heat and

* The architects of whose planning regime are, like the German Greens, invited to join the pro-catastrophe league.

† Both of these are isotopes of hydrogen. This means that they are structurally the same as a normal hydrogen atom, except that they have at least one neutron. Deuterium has one, tritium two.

the atoms fuse to form helium. As they do so, they release enough energy to either level a city or power it. The reaction takes place within a sausage-shaped mass of superheated plasma, kept stable with magnets. It is fiendishly hard to keep that superheated sausage stable, but if we manage to do it for a sustained period – and we are getting closer – then we will be on the brink of the greatest revolution in energy production since the invention of the coal-fired power plant.

Some of the most impressive progress towards achieving sustainable fusion has been made by British scientists.[44] Critics of the rush to net zero often point out that the UK's emissions constitute only a small portion of the world's, but it is through winning 'big bets' that we can have outsize impact. As Ackva explained: 'That's something governments can do – like the government of the UK, for example, which is ultimately a small part of global emissions and the global climate effort. Actions like this' – investing in technologies that might ultimately become alternatives to fossil fuels – 'can make a large difference.'

And, if we are lucky, humanity will win not one of those bets, but several. A pleasing side-effect would be the additional diversification of our energy supply, which is currently vulnerable to disruptions such as the Ukraine crisis. As a general principle, having more power stations, and more *kinds* of power stations, makes us more resilient. It might be that a few power stations get knocked out in a solar storm, or worse; or that some key resource is withheld by an autocrat. In either scenario, a wide array of power sources will help us ride out the instability.

Yet the rewards of a successful energy transition could go beyond mere resilience. As pointed out by the engineer Austin Vernon and the economist Eli Dourado, American innovation seems to have slowed over the period from 1979, in which the country has reduced its use of energy per person. As such, a good deal of our species' engineering and policymaking ingenuity has been directed towards efficiency rather than abundance. In short: we were promised flying cars, but we got regenerative braking.

But if one of those big bets yields us cheaper and more plentiful energy than we extract from fossil fuels, then the 'Great Stagnation', as the economist Tyler Cowen has called it, might come to an end. We can hope that the Great Stagnation will be succeeded by an age of 'superabundant energy', wherein we use at least ten times as much energy as we do currently. Vernon and Dourado envisage a world in which you might take a light aircraft to work, taking off from a local airfield as easily as catching a bus. Perhaps you would eat a lunch of cultured meat grown in a bioreactor. (Once too energy-intensive to catch on, this technology will allow humanity to end factory farming.) Your plane could be supersonic, bearing you swiftly and luxuriously to a land that was once an inhospitable desert, but is now a lush oasis. The oasis owes its transformation to the easy desalination of saltwater (again, hitherto overly energy-intensive) and the enabling of large-scale outdoor air-conditioning. In the same vein, colder countries could warm their city centres with outdoor heating. We could make a Scottish winter bearable or an Arabian summer temperate. We could irrigate deserts, bringing life and greenery to arid wilderness. 'Nevada would boom,' muses Cowen in his own consideration of superabundant energy, 'though a vigorous environmental debate might ensue: Just how many deserts should we keep around?'[45]

Cost is dependent on energy prices, meaning that almost everything would be cheaper. This would go for the manufacture of steel and concrete, giving us greater power to build new towns and cities that could, in turn, make us even wealthier.*

* The manufacture of steel and concrete together contribute an unholy share of humanity's greenhouse gas emissions. Newer production methods result in lower emissions.[46] I expect this trend toward lower emissions to continue by default and to accelerate if energy becomes a lot cheaper.

Perhaps humanity would invest its additional wealth into better healthcare, or more leisure time, or both. We cannot predict how exactly humanity would cash in its superabundant energy, but we can expect all manner of benefits. As Vernon and Dourado put it: 'Our belief is that, by the standards of the future, our energy consumption today, and therefore our standard of living, remains abysmally low.'

There remains, though, the matter of the tens of billions of tonnes of greenhouse gases that we dump into the atmosphere each year.[47] And those existing emissions have yet to take their full effect. Humanity's endeavour to avoid climate calamity, therefore, is a quest of two parts. We must rapidly rewire our entire energy system, but we must also clean up the mess that we have so lavishly sprayed into the sky. To do so, we must master carbon removal, via which carbon dioxide – the most abundant greenhouse gas – is scrubbed from the air.

At this point, one might well observe: *Congratulations. You have invented the tree.* But trees alone will not be enough. In order to compensate for a mere half of the emissions we have produced since 1960, humanity would have to reforest an area the size of Canada and the United States put together.* Perhaps simpler than depopulating an entire continent would be the creation of a technology that can do the trees' job for us.

And we have the technology: direct air capture. An existing carbon-removal plant in Iceland resembles a huge stack of air-conditioning units.[48] Each of these units sucks in air, filtering out the carbon dioxide. The carbon dioxide is then removed from the filter by heating it to a high degree. Then, with the addition of water, it is turned into a slurry that is disposed of harmlessly. The slurry is pumped deep into the Icelandic bedrock, never to be seen again.

* Such was the dispiriting verdict of an analysis published in 2009 in the journal *Science*.[51]

Governments are interested in this nascent method. Joe Biden's administration invested more than $100m in it.[49] But because direct air capture typically requires heat, it is not only expensive, but energy-intensive. All the more reason to make those big bets.

But we must ensure that carbon removal is feasible on a large scale even without an energy revolution. As the climate technology analyst Ben James observes,* a new process that centres on electrochemistry is an exciting new development. The method, he writes, is 'almost magical'. You run a current through some metal plates, thus capturing passing particles of carbon dioxide. 'Only electricity is required – no heat. This is hardcore science, still in development, and unproven at scale.'[50]

And perhaps we can take not only pollution from the air, but energy. An American startup, Terraform Industries, demonstrated in March 2024 that it could produce synthetic natural gas from sunlight, water and airborne carbon dioxide.[52] Via newly cheapened solar power, the company extracts hydrogen from water, adding carbon dioxide to create the simple hydrocarbon that is methane. Methane is the principal component of natural gas and it is a precursor to the longer-chain hydrocarbons that we use to create aviation fuel and rocket fuel. In other words, this prototype can conjure some of the world's best fuels, considered on their own merits, from almost nothing. It can plug into existing infrastructure, easing what might otherwise be a painful transition, and could allow us to stop drilling for more hydrocarbons. Its widespread adoption would allow us to re-use existing methane rather than pumping ever more into the atmosphere – thus saving us mammoth amounts of emissions.

* Ben is a friend of mine. Do not confuse him for an armchair environmentalist. He has charred the roof of the UK's leading start-up accelerator with a carbon-capturing limestone kiln that he knocked up with a small team of fellow enthusiasts.

In an ideal world, one in which we have minimised air pollution, we would not put captured greenhouse gases straight back into the air. But if we recycle them, we do not need to keep adding more to the atmosphere. Along the way, we prevent the six million tonnes of methane leakages for which, each year, the American oil and gas industry is responsible.* Terraform's mission is now to bring down the cost of its process such that natural gas that is taken from the air is as cheap as natural gas that is taken from the ground.

The company's founder, the Australian-born inventor Casey Handmer, is formerly of NASA's Jet Propulsion Laboratory† and of a project to build a hyperloop. (Hyperloops are a partially realised transport method by which cargoes – and, perhaps, one day, people – are sent zooming through the vacuums of airtight tubes.) Handmer is the kind of engineer who cannot consider a grand challenge without attempting to come up with an ingenious solution. In 2023, I interviewed him about his idea to make Mars habitable using hundreds of millions of reflective sails mounted on tiny, solar-powered spacecraft.[53]

These sails would reflect the light of the Sun onto this cold, arid planet. In doing so, it would raise the planet's temperature, thickening its atmosphere and making it possible for water to flow on the surface without freezing. Handmer's idea is a method of 'terraforming' – the creation of an Earth-like environment beyond our home planet.

Hence the name of Handmer's start-up, the work of which might contribute to the maintenance of an Earth-like Earth. 'Ten years ago,' he said in 2024,[54]‡ 'my colleagues and I looked at the

* This constitutes somewhere between 1 per cent of the oil and gas industry's methane production (not very exciting) and 10 per cent (much more exciting).

† To attentive readers, a familiar name. The Jet Propulsion Laboratory is the institution from which emerged the controversial proposal to prevent super-eruptions by studding calderas with geothermal wells.

‡ At the point Handmer refers to, he was a graduate student at California

prognosis for climate change, and it looked pretty hopeless. There really was no way out. You either continue to burn fossil fuels, and people have a decent quality of life, or you turn off all the oil and gas wells and your life regresses to something worse than North Korea extremely quickly.'

And then solar power became so cheap that even the most optimistic predictions turned out to be decades behind. 'Right now,' said Handmer, 'we're just beginning to feel the effects of increased rate of warming, increased intensity of storms, increased rate of melting of ice caps and so on. Now is the time to keep a lid on it. Now is the time to shove the genie back in the bottle with the technology we have at hand.'

Handmer's view is that climate change is to our generation as the Second World War was to his grandparents' generation. 'It was not what they would have picked for themselves, but it was what fell in their lap. They did a pretty good job with what they had. And we'd better damn well do a pretty good job of what we have.'

Innovators are rising to the challenge. In terms of manpower, the climate crisis is by far the most well-resourced of any of the anti-catastrophic endeavours I have reported on. New ideas abound. Mass timber, which is a wood-based product whose layers of glue make it less flammable, is proving to be a promising and environmentally friendly building material. (In 2023, I attended the opening of London's tallest mass timber building – a carbon sink that is as pleasing to the hand as it is to the eye.[55]) A British start-up has signed an agreement to beam solar power from space to Iceland's power grid, raising the tantalising possibility of harnessing sunlight before it is impeded by our atmosphere.[56] Electric batteries can now power light aircraft.[57] One day, hydrogen fuel cells might power passenger jets, as well

Institute of Technology. Here he examined the propagation of gravitational waves from phenomena such as the mergers of binary black holes.

as lorries and large boats.* Hydrogen could also help decarbonise steelmaking. As Molly McHugh, an investor in climate technology, told me: 'We have so many quick wins in front of us.'

We can predict, with some confidence, that fossil fuels will be thoroughly outmoded over the course of the 21st century. It is important that governments continue to underwrite those 'big bets': fusion technology, for instance, stands a much better chance of working if, through being given public funding, it is given the gift of time. Once a technology outperforms fossil fuels, though, it will need no subsidy. We will need no carbon taxes, nor carbon credits. Thanks to far-sighted government support, the market has greedily adopted solar power and electric vehicles.† Further greedy adoptions are inevitable. Bettering the miracle of fossil fuels is a lofty goal, but one that humanity is likely to meet.

That still leaves a serious problem unsolved. That problem is the warming (1.17°C/2.11°F) that has already occurred – and the warming that we are due from the greenhouse gases already in the atmosphere. Many people are trying to prevent the temperature from rising in the future; very few are trying to reverse the heating that has already occurred. But a small minority is attempting to do just that. Earlier, I referred to environmental dark magic; these are the environmental dark wizards. In 2024, I spoke to one of them: Andrew Song.

Song, a giggling iconoclast, rejects the worn pieties of climate

* In these fuel cells, hydrogen reacts with oxygen in a battery-like environment to yield electricity, water and a small amount of heat. Hydrogen is expensive to produce without generating excessive emissions and requires compression in order to be energy-dense – making it a relatively cumbersome option that is inferior, in some respects, to battery power. But hydrogen could turn out to be a useful option for large vehicles and ships.

† EVs are still generally subsidised by governments, but the subsidies will be phased out if and when costs continue to fall.[58]

change. 'A lot of people were indoctrinated,' he told me, 'including myself. I was an American boy scout, thinking: *I just need to recycle and recycle and plant some trees, and I'm good.*'

But these soothing instructions, Song came to realise, were 'all a fucking lie!'. Telling me the story, he burst into laughter. (And it's true – recycling is, in some respects, overrated.* It will not solve the climate crisis.)

Song went on to help start several businesses in San Francisco. While enrolled in Y-Combinator, the prestigious accelerator for start-ups, Song met Luke Iseman. This was in 2016. At the time, Iseman worked at Y-Combinator, but eventually he started pitching start-up ideas to Song. 'He would float crazy ideas with me,' Song recalled, 'saying, like, "Hey, what about this cat bed? Let's try to do a Kickstarter on cat beds."'

'And I'm like: "I'm not really a cat guy. Plus, I don't want to spend my time on this." But then, once, he said: "Andrew, I've just learnt about stratospheric aerosol injection."'

Stratospheric aerosol injection is the method by which we might be able to reduce the heat of the planet. It is a form of solar geoengineering. Recall the way that volcanic eruptions can blight summers or cause worldwide winters. They do so by launching sulphur dioxide into the stratosphere. Here, sulphur dioxide reacts with other gases to form sulphate aerosols, which are fine particles. These fine particles reflect the Sun's light back into space. And the more light is reflected back into space, the less it warms Earth.

It was in 1974 when a Soviet climatologist, Mikhail Budyko, suggested that global warming could be reversed if humanity could drape Earth's atmosphere in particles of this kind. The method was nicknamed 'Budyko's Blanket'. Seventeen years later, the eruption of Mount Pinatubo, in the Philippines, yielded further evidence that stratospheric aerosol injection might work.

* It's not particularly energy-efficient, for one. Don't imagine that recycling remotely compensates for your environmental crimes.[59]

The eruption took the lives of Maurice and Katia Krafft, the volcanologists I introduced earlier, but it demonstrated that even a VEI6 eruption could cause a drop in global temperature of perhaps 0.5°C.[60]*

At the time of the eruption, Song was six years old. A little more than 30 years later, he would be trying to emulate it. The intention, he told me, was: 'Let's try to replicate what Pinatubo did.'

Iseman had come across the idea in a science fiction novel. As he explained in an interview in 2024, he read widely on the topic and spoke to experts.[62] Something felt amiss. 'I was like, "Why has no one done this? What am I missing here?". No intellectually honest person debates whether climate change is real, yet we have this super-powerful tool and no-one has even done a small research release of the particles.'

Scientists had tried. Some researchers at Harvard, for instance, wanted to study the phenomenon by releasing a balloon, bearing an ingenious device, into the stratosphere.[63] Here the device would release some particles before measuring how much sunlight they reflected.

It was a project designed to gather knowledge rather than emulate the full effects of volcanic eruption. Yet it caused something of an earthquake. Having failed to find an appropriate launch site in the US, the scientists chose one in Sweden. There, in 2021, they were assailed by objections from environmental groups and a council representing the indigenous Sami people. In a letter that cc'd not only the Swedish Space Corporation (the host of the proposed launch), but also the government of Sweden, the scientists were told that there was 'no acceptable reason' for conducting the project.[64] The protestors warned of 'catastrophic consequences, including the impact of uncontrolled termination, and irreversible sociopolitical effects that

* A more recent study had the temperature reduction at 0.2°C – still significant.[61]

could compromise the world's necessary efforts to achieve zero-carbon societies'. The project was cancelled and the device was never launched.

The term 'uncontrolled termination' describes the scenario in which humanity starts a massive project of stratospheric aerosol injection, then abruptly stops.* It is thought that this would result in 'termination shock', whereby the temperature would swiftly rebound. Such a rebound would result in much more disruption of weather and ecology than would a more gradual rise in temperatures.

And termination shock is far from the only reason to be wary of stratospheric aerosol injection. One of those reasons is that it could deplete the ozone layer, which was damaged by 20th-century human-generated emissions, and partly healed as a result of a treaty so successful† that members of the anti-catastrophe league still cite it as a case study in international cooperation.[65] In addition, stratospheric aerosol injection could cause acid rain. It could have various other unintended effects, some of which we might not be expecting. And it would not address air pollution, acidification of the oceans and other nasty effects of the burning of fossil fuels. Instead, stratospheric aerosol injection might encourage us to stick with our bad habits.

Solomon Hsiang, an associate professor at the Goldman School of Public Policy at Berkeley, put it like this. 'We know that chemotherapy is a therapy that helps aid individuals with cancer, but it also has very traumatic side effects. There are some cases in which those risks are judged to be worth the consequences in order to avoid potentially larger harms.'[66]

Yet we continue to burn fossil fuels and the planet continues to warm. We have placed the big bets, but they have yet to pay off. And so stratospheric aerosol injection has become increas-

* Why might we stop? Perhaps another catastrophe, such as nuclear war.

† This treaty is the Montreal Protocol, which phased out humanity's use of chemicals that were responsible for the ozone depletion.

ingly attractive – the forbidden fruit in the garden of climate technology. The US government is developing a research plan for solar geoengineering.[67] It was followed by ARIA, the British government's innovation agency, which is also looking into the technology.[68]

By the time ARIA had announced its plan, the forbidden fruit had already been plucked. As Iseman later explained, 'I bought a couple hundred dollars of balloons on Amazon, a tank of helium and some sulphur powder, and we were off to the races.'

In April 2022, Iseman launched his first balloons, releasing them from his home in Baja California, Mexico. By the following February, *TIME* magazine was calling him 'an innovator, renegade, or charlatan, depending on who you ask, but certainly the biggest climate tech trouble-maker in recent memory'.

He and Song incorporated their new company, Make Sunsets, that October. (It is so-called for another effect of stratospheric aerosol injection: colourful sunsets.) Having been ticked off by the government of Mexico, they currently operate a patch of hilly Californian scrubland owned by Iseman. Journalists tagging along with Iseman and Song have witnessed a home-brew operation that resembles some of the more chaotic scenes from the TV series *Breaking Bad*: two rascals taking an RV to the wilderness in order to cook up something that most observers would probably call deeply irresponsible.

At the time of writing, the two men have launched more than 100 balloons. The balloons rise, rise and then pop, as verified by electronic instrumentation, expelling their sulphurous payload. The rest of the craft falls to Earth. The balloons are biodegradable, made of the same kind of material that weather services use for forecasting, and the instrumentation is trackable via GPS.

Song, a talented marketer, tends to liken stratospheric aerosol injection to sunscreen. As he pointed out to me, humanity already puts 60 million tonnes of sulphur dioxide into the troposphere each year.[69] (The troposphere is the lower few miles of

Earth's atmosphere; the sulphur dioxide comes largely from the burning of fossil fuels at power plants and industrial facilities.[70]) Song and Iseman propose simply to put a much smaller amount of sulphur dioxide a little higher up. In Song's view, we have been misapplying our sunscreen. We have been 'just spraying the sunscreen lotion on our face, but with our mouths open. And since some of that is being swallowed, we're getting sick off it.'

He characterised the Make Sunsets position as: 'Let's just close our mouths. That's all we've got to do, and then it won't go in our lungs. Put it a little higher, put it in the right places, and we're good.'

As for the ozone layer, Iseman and Song refer to evidence that its depletion will be comparatively small.[71] As for acid rain, Song accepts there is risk, but argues that the risks of inaction are even greater.[72] His view is that we should be less precious about geoengineering. 'We've injected two trillion tonnes of greenhouse gases into our atmosphere since the 1870s,' he told me, 'so we are we suddenly drawing a line of where we can deploy these aerosols? Literally, as we speak, we're emitting CO_2, geoengineering the Earth, even just by breathing.'

Convinced? For $10, you can buy a Make Sunsets 'cooling credit' that will offset, their website states, 'one ton of CO_2 warming for a year. The average American generates 16 tons of CO_2 per year.'[73] Iseman and Song estimate that all man-made global warming could be offset for $50bn per year. They have been told to keep various US government agencies apprised of their activities, but they have not been told to stop.

At least not by governments. Technologists such as Casey Handmer are supportive, but there has been no shortage of criticism from academics and environmentalists. 'I have not seen a single person in the field say this is a good idea,' a lecturer in environmental studies told *TIME*. Song thinks the critics are blind to the planet's current predicament, but agrees with them on at least one thing to do with solar geoengineering. He and Iseman, he told me, 'should not be the ones doing this. Like, to be very clear about this, we should not be the ones doing this!'

Who should? Ideally, there would be international consensus. That's what happened with the ozone layer. But consensus takes a long time to build. A non-governmental organisation, the Degrees Initiative, has been working for a decade to help developing countries work out how enthusiastic they are about solar geoengineering. Decades-long deliberation processes might be appropriate for ideas such as volcano manipulation; you will recall such a plan from the chapter on super-eruptions. But the climate crisis is urgent. And the longer one thinks about solar geoengineering, the harder it is to avoid the conclusion that climate change – as we know from the needless discarding of nuclear power – is, to a meaningful extent, a choice.

My prediction is that solar geoengineering will, without much warning, become a mainstream idea. It could happen within the next year or two. Consensus is not unanimity, though; it seems unlikely that solar geoengineering will have wholesale worldwide support. How many countries need to consent for others to go ahead? What if changing weather patterns disadvantage one country more than another? In those scenarios, who compensates whom? Will countries participating in a stratospheric aerosol injection programme use it as an excuse to withdraw from the Paris agreement? The quick resolution of these competing problems will be a miracle of its own – one that might give us insight into how best to deal with other fast-emerging new problems, such as misaligned or misused artificial intelligence. Perhaps it is naïve to imagine that solar geoengineering will occur after diplomatic efforts rather than before. As Ben James writes: 'I find it impossible to imagine a UN mechanism approving something so universally contentious. Rather, someone will probably just do it.'

Where the climate crisis differs from other perils, however, is that it is something we all contribute to. It has therefore provoked much worry, guilt and overzealous recycling. Perhaps the most pragmatic thing an individual can do about it, aside from holding off on flying to Paris for cheesecake, is donate to organisations that are trying to accelerate the development and adoption of clean-energy technology. One can do this via Ackva's climate change fund, housed within a charity called Founders Pledge. The non-profit Giving Green operates a fund that is based on similar pragmatism.* (And there is nothing stopping you from paying Iseman and Song to release some sulphur dioxide on your behalf.)

If humanity is to rescue itself from the worst effects of climate change, it will not be because our species acquired a taste for hairshirts. It will be because of thick-skinned weirdos who raised the alarm, hard-headed innovators who created products that the market prefers to fossil fuels, and crafty policymakers who made big bets that paid off. (Perhaps it might also be because of some balloon-releasing cowboys.) Should things go particularly well, this anti-catastrophic coalition might help us reach a state of energy superabundance.

There is one more big bet, though, that I have hardly mentioned. Many discussions of clean energy omit this technology entirely. But these omissions, I believe, are misguided. In the next chapter, I will reveal how an ancient and powerful energy source might be the best bet of all.

* Its recommendations include organisations that address the emissions of heavy industry, aviation and livestock production, while spurring the adoption of clean-energy sources.

CHAPTER 7
SUPERCRITICAL

'Before their eyes in sudden view appear
The secrets of the hoary deep ...'

John Milton, 'Paradise Lost'

The deepest hole humanity has ever dug is the Kola Superdeep Borehole. Soviet scientists, hoping to learn about the composition and geophysics of Earth's crust and upper mantle, began drilling on the Kola Peninsula, in the country's frozen northwest, in 1970. After more than two decades of tearing into the bedrock, they reached a depth of 12 kilometres – a distance 113 times the depth of the world's deepest metro station: Arsenalna, in Kyiv.

Along the way, the scientists found mineralised water deeper than any of them had thought possible.[1] They found microscopic plankton fossils that were more than two billion years old. And they pressed on. At 12 kilometres deep, their slender borehole was less than 0.2 per cent of the way to Earth's core, the centre of which is thought to be 5,200 degrees Celsius. That temperature isn't far off being twice as hot as the temperature necessary to vaporise iron. Knowing this, the scientists expected the bottom of their 12-kilometre hole to be 100 degrees Celsius, hot enough to boil water. Instead, it was 180 degrees Celsius, the heat of an oven.

At those depths and that temperature, the drilling was pushing the limits of technological possibility. It's difficult to get good performance out of a drill bit that far from the surface. It's hard to dispose of 'cuttings' – the rock torn up by the drill – and harder still to keep replacing those drill bits, which quickly get gnarled up, without the swapping process becoming increasingly time-consuming as the wellbore deepens. In these extreme conditions, granite behaves more like plastic than rock, creating further difficulties. And the heat is ruinous to the equipment that accompanies the drill bit. The hole became more and more challenging to drill and the project ran out of money.[2] In 1995, the nine-inch borehole was welded shut.

The borehole remains stoppered. But the promise remains that by digging deep into Earth, deeper than the Kola Superdeep Borehole, we could bring to the surface clean, reliable energy in vast quantities. Nuclear fission, so far, has been kneecapped by regulators;[3] renewables have made great progress, but owing to their intermittency, they cannot currently solve our energy requirements; fusion continues to elude us. Energy is still hugely important for human advancement and a lack of it might help explain our society's post-1970s economic stagnation.[4] How might we not only meet our current demands for energy but exceed them? The answer could be under our noses.

At the same time as the Soviets were drilling the Kola Superdeep Borehole, Paul Woskov was trying to crack nuclear fusion. He'd been trying for some time. He was 20 years into his career as a senior research engineer at Massachusetts Institute of Technology, but the holy grail – endless clean energy – seemed little closer. The fusion reaction, you will recall, occurs within superheated plasma. Woskov heated that plasma by firing millimetre waves at it. Within the electromagnetic spectrum, millimetre waves lie between the super-high frequency band and the far infra-red band. Classified as 'extremely high frequency', they have wavelengths from one to ten millimetres, hence the name. Millimetre waves, in other words, are high-powered beams of energy. They are similar to the wavelengths commonly used in telecommunications, but of a higher intensity.

To produce these beams, Woskov's lab used a gyrotron, which is a complex assembly of Soviet-invented machinery. Gyrotrons use a strong magnetic field to create highly concentrated millimetre waves. Gyrotrons are costly to build, costly to run and difficult to maintain. But they are the best technology available for heating plasma to the 100 million degrees Celsius that is necessary for the fusion reaction to occur on Earth. (The Sun has the slightly unfair advantage of having a huge gravitational field that holds the ionised gas together, hence achieving fusion at a balmy 15 million degrees Celsius.)

In 2008, when an MIT energy initiative announced that it was looking for technology that would improve drilling, Woskov realised he might have an answer. 'I knew that if you bring enough energy to bear on matter, matter is destroyed,' he told me in 2022. His speech was leavened by deep chuckles, but was mainly composed of the unrelenting technical matter that is emitted only by career scientists.

As Woskov explained, the most common types of bedrock are basalt and granite, so he brought a lump of each into the lab. When he heated a sample in the gyrotron's test chamber, he heard the rock fracturing. Fragments of rock would sometimes rattle around the chamber, and the process often created the smell of hydrogen sulphide: quick-dissipating, but eggy.

The fracturing of the rock opened an intriguing possibility. It has been known since experiments with lasers in the 1990s that drilling with heat could be much faster than conventional drilling. But lasers quickly get neutered by the dust produced by the drilling process, which absorbs so much light that the lasers can't get past it. The microwaves produced by a gyrotron, on the other hand, can be guided for kilometres and are unaffected by dust, making very deep wells possible. And those deep wells could unlock a power source as exciting as fusion.

It is the fuel source that has been under our noses for as long as humanity has existed. Long before we arose, some 4.5 billion years ago, Earth was formed by the collision, accretion and compression of space rocks. The residual heat from this process, alongside the heat produced by the decay of radioactive elements, such as thorium, potassium and uranium,[5] is the ultimate source of the furnace-like temperature of the planetary core. The core accounts for 15 per cent of Earth's volume, but – as a result of the hellish pressure of the weight of a planet – 30 per cent of its mass.

Our planet's innards, then, are infernally hot. Sometimes we encounter those innards directly, when they are disgorged by a volcano. Sometimes we encounter them at a more amenable distance. Roman bath houses made use of hot springs near volcanoes, where magma is close to Earth's surface. In drizzly southwest England, the hot springs in Bath in Somerset, atop which the Romans built a complex of temples, derive their heat from deep within Earth's crust. Rainwater percolates down through limestone aquifers to as deep as nine kilometres and is heated by the hot Earth. Pressurised, the water is forced back upward, picking up minerals from the limestone as it goes. But the Romans were late to this particular hot-tub party. The archaeological record suggests that ancient hominids were using hot springs more than 1.8 million years ago.[6] We can safely assume that humans have always enjoyed a hot bath.*

It's one thing to splash around in preheated water; it's another to turn that heat into electricity. In the early twentieth century, it might have looked that geothermal power was on an exciting trajectory. In 1904, Piero Ginori Conti, an Italian who ran a boric acid extraction firm, used the natural dry steam geysers of

* Sometimes the bath can be too hot. In Lassen Volcanic National Park, a Californian wilderness I visited in 2024, there is an area of hot springs and mud pots that is known as Bumpass Hell. It is named, alas, for Kendall Vanbook Bumpass, who fell in and lost a leg.

his native Tuscany to power a dynamo – an early generator – that lit five lightbulbs. (Dry steam, as opposed to wet steam, is so hot that it doesn't carry any water droplets in it: it is 100 per cent gas, rather than partly liquid.)

Larderello, the Tuscan village where Conti lit those lightbulbs, still produces electrical power from geothermal energy. The dry steam geysers of California, which are similar to Larderello's, have supported power plants since 1960. (In 2022, they produced five per cent of California's in-state electricity generation,[7] doing so by pumping water into the ground and using the steam that subsequently arises to turn a turbine, which drives a generator and, ultimately, creates electricity.) After the US, the present day's biggest producers of geothermal energy are Indonesia, the Philippines, Turkey and New Zealand.[8] Kenya and Iceland produce less in total, but they get most of their electricity from geothermal plants.

The long history of geothermal power, and its wide geographical spread, might seem to suggest that humanity has made progress in exploiting it. But that is not the case. Each of the countries I have mentioned are taking advantage of quirks of geology. They sit at the meeting points of tectonic plates, which allows them to access geothermal heat (and thus power) without massive drilling. These circumstances also explain why each of these countries is speckled with volcanoes. To change the world, geothermal power must be accessible far beyond these specific regions of volcanism. It must be accessible in places as geologically boring – and I say this with love – as Britain.

In total, geothermal contributed 15.7 gigawatts of worldwide electrical power in 2020 – enough to power 200 million lightbulbs.[9] It is an amount that Conti, the Tuscan inventor, could hardly have dreamed of, yet it is a sliver of global energy production: an estimated 0.35 per cent. This figure seems even more meagre when one considers that an infinitesimal fraction of Earth's geothermal heat – a tenth of one per cent of that heat, goes the calculation – would power humanity's current outgoings for 20 million years.

With no recourse to volcanoes, hot springs and fumaroles, developers are left with challenging options, such as deep drilling. As fans of the Kola Superdeep Borehole will know, deep drilling requires huge upfront spending, with no guarantee of success. And it is extremely technically challenging – not only in terms of the drilling itself, but also in terms of working out where to drill.

As well as that, geothermal projects have a record of causing tremors. These tremors are caused not by the drilling itself, but by the use of high-pressure fluids to force open underground pathways. (Most commonly used to extract shale gas, this is similar to the hydraulic fracturing used by the oil and gas industry and better known as 'fracking', but at lower pressure.[10]) The pathways create a network through which water can pass, allowing engineers to pump it down through one well, then draw the same water, now heated, back up a neighbouring well. But this process can cause tremors. No wonder geothermal is unpopular. This combination of hurdles – the cost, the uncertainty, the risks – have made geothermal projects much more difficult and expensive than drilling for gas.

There was a crack of hope for geothermal in the 1970s. A worldwide energy crisis compelled governments to look into alternative power sources, prompting brief interest in geothermal. In the end, though, fossil fuels retained their primacy. Even in the race to net zero, governments and environmentalists have preferred wind and solar energy. The world's production of geothermal power has been inching upward rather than rocketing.[11] It has been a dribble rather than a geyser.

Woskov, then, had to show that the millimetre wave could make the difference. His proof of concept was diverting, but whether it was meaningful would depend on its cost. The method had to make for cheaper energy than, in oil and gas, the greatest fuels of all time.

First, Woskov had to find how much heat was needed to reduce those rocks to fine dust. This was a difficult number to identify, because not many people had tried it before. But according to Woskov's calculations, the process would be many times cheaper than the conventional method. Where a mechanically drilled eight-kilometre well typically costs about $50 million to drill, to create a similar well through the gyrotron method would be much cheaper. To do so using millimetre-wave-based technology, Woskov later told me, would use between 916 and 5,750 megawatt-hours of electricity, equivalent to 11,000–70,000 charges of a Tesla Model 3. At $120, or £94, per megawatt-hour (which is about how much electricity costs in Texas),[12] this would only cost between $110,000 and $690,000, (£86,000 and £538,000) – depending on whether the rock was melted, which would be cheaper, or vaporised, which would be more convenient, but more energy-intensive. While this estimate excludes labour and capital costs, Woskov says these would be much lower than in conventional drilling, owing to the simplified process.

In that same year, 2008, Woskov and his MIT colleague Daniel Cohn filed a patent for what they called a millimetre-wave drilling system. There were many technical challenges to overcome, notably those concerning this novel use of gyrotrons. In principle, a powerful gyrotron should be able to fire continuously for 24 hours a day, seven days a week – surpassing the current record of an hour at a time for an 800 kilowatt model, and eight continuous hours for a 200 kilowatt model.[13] But to direct that power several kilometres into the ground was an endeavour never before attempted, let alone successfully executed.

And successful execution – deploying a technology so well that it becomes commercially viable – was a challenge Woskov could not overcome alone. Enter Carlos Araque, who had been working in the oil and gas industry for more than a decade when he felt what he calls an 'inkling of the soul'. Araque, who grew up in Medellín, Colombia, had joined Schlumberger, the world's largest offshore drilling company, after graduating from MIT with a master's degree in mechanical engineering. (He did not meet Paul Woskov at MIT, but their paths would cross in the years to come.) He told me that his career had been 'very comfortable': it paid well, it was technically interesting and it provided the stability that helped Araque support his family. He and his wife, Tina, a business consultant whom Araque met at MIT, have three children, of which the eldest, Alejandro, is physically disabled from birth. But Araque could not ignore that inkling. 'Coming of age,' he told me, 'coming closer to my forties, I said: "Okay, the world needs to transition its energy. That's just a fact. I'm convinced of that."' He asked himself two questions: 'How can I help?' and 'Where can I help?'

There were many reasons to ignore the inkling: bills, college fees, logistics, and so on. Araque's parents advised him to stay at Schlumberger. But, as Araque would later tell me, 'life ends. You have to do something that matters, and you have to do something that you're convinced by and proud of.' With his family now supporting his decision, he quit his job. He had no new role lined up, just an urge to find some way of accelerating humanity's transition away from oil and gas, to which he had devoted his whole career, and towards clean energy. So much for his worries. 'Once you're able to overcome all this bullshit,' he told me, 'all these excuses that you give yourself, then you break through and find your way.'

Araque returned to Cambridge, Massachusetts, to help set up The Engine, a venture capital fund founded by MIT. The fund invited MIT academics to come and pitch to them at 501 Massachusetts Avenue, Cambridge. On a sunny day in July 2017, Paul Woskov, encouraged by the investor Aaron Mandel, walked into room 3B, carrying the rocks he'd blasted those holes into. On the other side of the table was Araque. He was fascinated by the rocks, intrigued by Woskov's calculations.

For the next 15 months, Araque immersed himself in Woskov's work, reading everything he had published over the preceding decade. He spoke to everyone he knew who could help him understand the science that underlay the proposal. He had meeting after meeting with Woskov, who was as powerful a physics gyrotron then as he was when I spoke to him. 'Paul is a career scientist,' Araque told me, 'and when you ask him to simplify things, it's still at a very high level. Luckily, I'm very technical myself and I'm not shy to roll up my sleeves and really whiteboard these things out and write the equations.'

In these intense, private tutorials, Woskov taught Araque how gyrotrons work, how waveguides and transmission lines come into play, and how the hardware would translate into vaporised rock. Over this period, Araque was introduced by Mandel to Vinod Khosla, a Silicon Valley-based venture capitalist, making the case for investment in a company that would put Woskov's vision to the test. Khosla agreed, with one condition: that Araque be Woskov's CEO.

It was a proposal that suited both men. Woskov, the brilliant scientist who knew little about business but everything about gyrotrons, had found a technically minded CEO who believed in him. Araque, an oil and gas man who dreamed of consigning his old industry to history, had found a scientist whose idea could do just that. Recounting what he made of Araque in that first meeting, Woskov said: 'Oh, I thought he was a very energetic person. Very motivated. He realised the potential pretty early, and he's intelligent. So I had a favourable impression of him when I met him, and I still do.'

With Aaron Mandel, the entrepreneur who brought Woskov to the pitching room, and Matt Houde, a civil engineer and geologist, Araque and Woskov spun their new company out of MIT Plasma Science and Fusion Center in 2018. Mandel gave the company its name: Quaise. It is the name of a part of Nantucket, chosen for its memorability and for its echo of 'quasar', a kind of celestial body that emits more light than even galaxies. So powerful is this light that it can destroy stars and vaporise planets. 'We're not going to do that,' Araque told me firmly.

Quaise's plan is to use millimetre waves to vaporise bedrock, the solid rock that lies beneath soil. This method, if feasible, would make it possible to create holes as deep as 20 kilometres. At those depths, rock is 500 degrees Celsius: super-hot rock, as it's called. Pumping water into super-hot rock gives us supercritical water – water pressurised and heated beyond 373 degrees Celsius, at which point it has the low viscosity of a gas, the high density of a liquid and an energy density not far off from that of fossil fuels. Supercritical water is so energy-dense that a successful attempt to commercially harness it would be, in the words of the volcanologist and petrologist Mike Cassidy, a 'complete game-changer'.

Quaise claims that, by 2026, it will be able to produce 100 megawatts of thermal energy from 'a handful of wells'. Perhaps half of those megawatts, depending on the efficiency the company can achieve, would become usable electrical power. A typical American nuclear plant, for context, produces one gigawatt of energy (1,000 megawatts). But, in Quaise's vision, new wells could be created far more easily than new nuclear power stations, and in much greater numbers. They could even replace the gas- or coal-fired boilers of existing power plants.

Cracking millimetre-wave drilling, said Araque, would unlock all kinds of doors. It would enable us to access mineral and metal deposits that we can hardly dream of; to attempt new and ultra-ambitious forms of civic construction and tunnelling; to experiment with using Earth's gravity to send cargo, via tunnels, from one point on Earth's surface to another. Here is a man who has read his Jules Verne. 'This technology will open up frontiers that are not imaginable,' he said.

In the first week of March 2023, Araque's team hit a significant milestone. With the help of a millimetre wave, they vaporised a hole 100 times deeper than its diameter, reaching a depth of ten metres. It was a long way from 20 kilometres, but it was 100 times deeper than Woskov's earliest holes, and it was the landmark the team needed to reach before taking their work into the field.

Others began to believe in this vision. 'I talked to my director at the Plasma Science and Fusion Centre,' Woskov recalled. 'He's supportive and he realises the potential. He looks at me and says, "Paul – my God. You're going to compete with fusion!."' Woskov smiled as if even he could hardly believe it.

Is Quaise's plan feasible? Quite what it's like 20 kilometres down, even if they make it that far, is anyone's guess. 'We're going to a place nobody's ever been,' said Araque. 'How do you know what it's made of? It's like man touching down on the Moon for the first time.'

Experts I spoke to explained that Quaise would need to overcome a succession of technical challenges. There are the logistical challenges of setting up the drilling rig and obtaining the right gyrotron. (At the time I spoke to him, this was Woskov's quest.) As I've mentioned, this gyrotron will also need to run 24/7, with all the power transmission and cooling that that requires.

And these problems are – as it were – surface-level. Millimetre waves require a high concentration of energy: how would that energy be transmitted several kilometres down the wellbore? And how would the technique work in unfamiliar subterranean territory? High-temperature rock tends to perform in a ductile rather than a brittle fashion, meaning that it slowly shifts rather than staying rigid and snapping. Would the wellbore close over time? What happens when bedrock vaporises? Does it condense back into ash, falling back down the wellbore? A literature review turned up nothing, Araque told me. 'Nobody has gone to vaporise rocks for a living.'

The most difficult challenge of all, several experts said, was the heat. The temperatures that make Quaise exciting are temperatures far beyond what any existing drilling equipment can withstand. All modern drilling involves downhole electronics that monitor what's going on at the bottom of the hole. Subjected to intense heat, those electronics rapidly lose battery life. Then they melt – and so does the well's casing. Araque wants to reach rock that is 500°C, but today's top-of-the-range equipment can barely withstand 200°C. The Quaise project, an industry veteran told me, is 'snake oil'.

The Quaise view is that the oil and gas industry hasn't had to deal with temperatures that high and so we shouldn't take the absence of those tools as evidence of their impossibility. Investors seem to agree.

Nevertheless, the scepticism is reasonable – difficult though it might be for geothermal fanboys to digest. I am one of these fanboys; I long for one weird trick to unlock the power of Earth's core. After writing about the geothermal sector as a journalist, I studied it on behalf of a think tank, spending weeks researching the best ways by which the US government can make the industry viable. The blood runs hot in my veins as I imagine supercritical fluid gushing through the pipes of converted oil and gas infrastructure. But even a swooning fanboy must acknowledge that the one weird trick might not work. Just as humanity is, quite rightly, betting on several clean-energy technologies at once, so we should be betting on several ways of extracting geothermal power.

The surest bet relies on the stunning progress made in the shale industry. Shale is a fine-grained sedimentary rock whose structure traps organic material. Over millennia, the organic material turns into hydrocarbons – those long, energy-rich chains of molecules whose synthesis I discussed in the previous chapter. For most of modern history, shale was less enticing to the oil and gas industry than were other fossil fuels. But from the 1990s onward, everything changed. For one, the oil and gas industry had exhausted many of the more easily accessible deposits of fossil fuels. Necessity, as ever, was the mother of innovation, and the industry worked out how to extract shale gas effectively.[14] One important innovation here was the invention of horizontal drilling, allowing drill apparatus to descend vertically before reaching new regions within shale deposits by travelling laterally. Another was the development of the fracking technique. Forms of this method already existed – hence the tremors that helped put people off geothermal – but the oil and gas industry could now frack at high pressure, reliably extracting cheap shale. The shale boom of the Noughties made the US the world's top oil and gas producer.

It is less sexy than using lasers. But the importing of these methods to geothermal is a promising new frontier, even if that frontier will advance incrementally rather than overnight. In 2014, the US Department of Energy announced its plan to create an unusual new institution: a $218m field laboratory whose remit was to conduct hands-on research into this fracking-inspired form of geothermal. The method is called enhanced geothermal systems, known for short as EGS. It would be explored in the desert of Nevada, with a small team of academics and engineers given a few football pitches' worth of scrubland. Under that area was an unusually hot subsurface, and around it was a sparsely populated county whose inhabitants were used to the mild disruption of local fracking projects. The funding was to last for ten years – ten years in which the field lab, FORGE,* was tasked with finding out how to make drilling so cheap that EGS could become commercially viable.† Only at that threshold would EGS start to become a workable solution for the climate crisis.

For geothermal drilling enthusiasts, this patch of scrubland was suddenly their equivalent of Disneyland: a magical realm where dreams, at least those pertaining to new drilling methods, could be made reality. The FORGE team tried particle drilling,

* Frontier Observatory for Research in Geothermal Energy.

† DOE's goal is to reduce the cost of EGS by 90 per cent, to $45 per megawatt hour by 2035. (Solar is $23 per megawatt-hour and a combined-cycle gas plant comes in at $43 per megawatt-hour.)

an experimental drilling method in which ball-bearings are fired at high velocity into the earth. (It didn't work, but was worth trying.) They tried the mud hammer method, where drilling fluid is used to power a piston motion that augments the rotary motion of the drill. (Ditto.) That these methods didn't work is no mark of shame; they are smaller examples of the 'big bet' thinking. Bet enough times on outcomes that are low-probability but high-reward, and eventually you'll get a big payout.

The supercritical enthusiasts note that FORGE was more preoccupied with high temperatures than drilling speed and reservoir management. (The reservoir is the subterranean network of cracks that heats the fluid that engineers pump through it. The bane of geothermal projects is the loss, over several years, of the reservoir's heat – a nasty surprise for investors.) FORGE should be drilling deeper, they say, reaching hotter rock. The FORGE team, led by Joseph Moore, point out that they are already working at temperatures of 220°C – as hot as the downhole electronics can handle. They are working with companies to improve the durability of these electronics; one of the great things about this government site is that it offers a low-risk environment for geothermal companies to test new equipment. 'Fantastic, right?' remarked Dr Philip Ball, of Geothermal Energy Advisors, when I spoke to him about FORGE. 'Who else offers you an 8,000-foot hole to put your tool down?' (If the high-tech work government project had made you think of the Manhattan Project, note that this is a key difference: Oppenheimer was not running a bring-your-uranium-235-to-work sort of operation.)

The year 2024 marked the end of FORGE's initial period of funding. Commissioned by a think tank, I spent much of that year examining the lab's performance. Thanks to factors such as inflation, Moore and team had had to spend far more on hiring drilling rigs than they had projected. Nevertheless, their drilling speed had increased fivefold – *fivefold!* – between 2017 and 2022. Drilling costs had fallen by as much as 50 per cent from

2022 onward. And because all of FORGE's research was public, the entire geothermal industry could benefit from it.

Those geothermal industry colleagues now know that a polycrystalline diamond compact drill bit, for instance, is a wise import from the oil and gas industry; thanks to its artificial diamond coating, the bit is able to cut through hard rock more efficiently than its rivals. They have a better understanding of how to keep reservoirs alive and roasting for longer and longer. And they know, as a result of FORGE's clever use of sensors, that sometimes drilling is slowed not by rock, but by a thick mixture of drilling fluid and rock cuttings. FORGE devised a crafty solution: flushing out the clog with a 100-gallon water 'pill', thus halving the drill bit's energy consumption and doubling its drilling rate.

Perhaps the most telling sign of Moore's success, though, is the arrival of a new neighbour: a geothermal start-up called Fervo. FORGE showed how to make that patch of Nevada scrubland into the site of viable wells, and now the lab's research has prompted a commercial imitator that has signed a deal to supply energy to a local utility company and to Google.[15] Fervo's chief technical officer confirmed to me that FORGE's research had been 'extremely valuable' to him and his team.

The real test of Fervo's quality will be whether they are providing cheap energy several years into the use of a reservoir. But FORGE seems to have made important contributions to geothermal and, in October 2024, its funding was renewed. This gives Moore another four years to drive forward the field of geothermal power. Experts hope to see work on high-temperature electronics, closed-loop systems – in which engineers drill a subterranean radiator-like structure, so as to better retain heat – and, of course, cheap drilling. We might not even need the millimetre wave.

In the best case, the costs will be low enough for the oil and gas sector to move in. On that topic, I spoke to Cindy Taff, formerly the vice president of Shell's unconventional wells operations and now the chief executive of Sage Geosystems. This is a Texan geothermal start-up that employs off-the-shelf oil and gas technology, as well as off-the-shelf oil and gas chief executives. On the day I spoke to Taff, her team had been running tests at an abandoned gas well. She said that the oil and gas sector, hitherto put off geothermal by the poor investment return relative to fossil fuels, was 'watching and waiting for the industry to crack the nut on making geothermal more cost-effective'.

In the previous chapter, I quoted depressing figures about the UK's energy prices. It is exceptionally hard to build power stations in this country, even those of the most simple variety. Yet geothermal is demonstrating potential even here, a land where volcanoes were last active 50 million years ago. For 2,000 years, those Roman ruins remained largely uncontested as the apex of national geothermal infrastructure. Cornwall, the county that forms the tip of the mainland UK's southwest peninsula, has, like the rest of the country, been untroubled by volcanoes since the Paleogene era, which was the interval between 66 million and 23 million years ago. To use the lingo, Cornwall has a relatively high geothermal gradient. This means that, as you go farther underground, it gets hotter faster than is typical. Drill a one-kilometre-deep borehole elsewhere in the UK, and the temperature at the bottom will be 20 degrees Celsius. Drill a hole of the same depth in Cornwall and the temperature will be close to 40 degrees Celsius. This is not because Cornwall has more exposure to Earth's core, but because the peninsula's granite backbone contains radioactive potassium, thorium and uranium. The bedrock acts something like a natural nuclear power plant. This makes Cornwall, a county of lush greenery and abandoned tin mines, one of the UK's more promising venues for harvesting geothermal energy.

The Cornish greenery is its lushest at the Eden Project, where two sets of latticed, transparent domes play greenhouse to thousands of species of plant. One of the sets of domes houses the largest indoor rainforest in the world; the other is home to a similarly lush Mediterranean ecosystem. The project, built in a reclaimed china clay pit, was opened in 2001, a time when humanity was rather less concerned about climate change than it is today. Eden, as it's known, needs around 3.5 gigawatt-hours of electrical power per year, and around five gigawatt-hours of heat. Powering it, Eden's managers decided, should be a geothermal energy project. The project broke ground in May 2021 and was switched on in June 2023.[16] Its 3.8-kilometer-deep heat main now delivers 85-degree-Celsius water to heat the greenhouses and offices of the Eden Project.[17] It would also be providing electricity, but it is expecting to wait 13 more years for a connection to the grid.

Gus Grand, who led Eden's drilling project, showed me the site late in 2022. Visible through the fence's lattice was a set of spent drill bits. Battered, rusty and lined up in a row, they looked like they had been left there as an example to any drill bits considering stepping out of line. The bits were designed for the oil and gas industry: three fat, studded wheels in a housing of chunky steel. The steel was now orange-brown. The studs, which are the teeth via which the drillers slowly grind through the granite, were made of ultra-tough tungsten carbide, but had been thoroughly worn down. When in use, each bit hangs off the end of a drill string that, in this case, had to go five kilometres deep. It's hot down there, it's wet, and it's kilometres away from the team on the surface. Grand spoke frankly and entertainingly about the challenges. 'If you drop the drill bit down the hole, you're really stuffed. You have to get a big magnet down from Aberdeen.'

All this goes to illustrate the technical challenge of deep drilling. It is not easy – and we're only just beginning to master it. Another challenge is the dearth of clarity over regulation from the Health and Safety Executive (HSE), the UK government agency that oversees occupational risk. Because the HSE does not have specific geothermal regulations, Eden had to default to full oil and gas regulation. As a result of its imprecision, this regulation cost the project money and time. For example, Grand's team had to hire, and regularly test, a 'blowout preventer' – a chunky piece of kit that prevents the uncontrolled release of natural gas, or crude oil, from within a well. The geothermal well contained neither natural gas nor crude oil, and the hiring of the blowout preventer and its operators cost the project, Grand said, tens of thousands of pounds per day. A similar lack of regulatory clarity affects the geothermal industry in many other places. In the US, for instance, many states have not yet clarified who owns the mineral rights to the heat accessed by deep drilling.

The challenge most immediately relevant to public opinion, however, is that of avoiding tremors. Grand's project targets an existing fault, and is gentler than the fracking practised by the oil and gas industry. Rather than blast rocks open, Eden needs merely to pump water through existing fissures. But at 9.20 p.m. on a still night in March 2022, people up to three miles away from the drill site felt their homes shake.

'It was a real bugger,' Grand said. A small, underground movement of water, during a pressure test, had triggered a 1.5-magnitude earthquake.[18] It wasn't big enough to cause any damage, but it did not escape locals' attention – though it is worth noting there are hundreds of similar sized or larger earthquakes across the UK each year.[19] Work was paused and Grand was hauled up to explain herself on BBC Cornwall. 'Because [the project] is a sort of greater-good thing, it was okay,' she told me. 'But I don't want to be doing that again.' There has been no repeat. The site, nestled in countryside, is quiet, compact and unobtrusive, taking up little more space than a couple of tennis courts. Alongside it is a well-trodden bridlepath. Before the work started, said Grand, 'we had everyone going: "Oh my God, the horses will be terrified by the drilling rig".' But the subterranean drilling produces little external noise. Grand said the project's loudest recorded sound is from the dawn chorus in the neighbouring wood, which is composed of oaks and willow.

It was a miniature marvel of a power station. Its running will be cheap but, at £24 million, its construction was expensive. The 11 gigawatt-hours of heat it can produce – the equivalent of heating about 900 homes – is therefore costly compared to conventional forms of power generation. Grand, however, sees Eden Geothermal as a research endeavour, one that will help the cost of geothermal follow a similar (rapidly downward) curve to the cost of wind and solar, ultimately becoming competitive with other forms of energy.

Cornish geothermal offers a tantalising glimpse into another potential upside of deep drilling. It is the final upside that I will describe, and it is perhaps even sexier than the millimetre wave. As context, know that the water which emerges from geothermal systems is called brine, and that Cornish brine is extremely high in lithium: 260-odd parts per million, reports United Downs, Cornwall's other ongoing geothermal project. The project's leadership has not yet disclosed the potential value of the lithium it stands to harvest, but its hope is that the metal will further strengthen the economic case for deep geothermal heat and power. Lithium is essential for the lithium-ion batteries that power green technology such as electric cars. Although there have been several recent discoveries of significant deposits,[20,21,22] it remains a metal in short global supply. But it is not the only metal that can be mined, as the terminology goes, from the brine.

Jonathan Blundy, a professor of earth sciences at the University of Oxford, would like to mine that brine from the flanks of dormant volcanoes. Lithium, gold, silver and manganese are among a broad portfolio of metals that lurk in the fluid extracted by deep geothermal. Sitting at his desk, with rocks lining the bookshelves, Blundy explained to me that magma and volcanoes are 'the main transport agent bringing these minerals from the Earth's interior to the surface'.

This means that geothermal projects, if sited by those dormant volcanoes, could bring up extremely valuable metals. Sprinklings of Cornish lithium are one thing; the treasures of volcanoes are another. Through such projects, Blundy said, we could defray a huge fraction of the cost of geothermal power, reducing the net cost 'by a factor of easily two, maybe ten'. This would not only offset the cost of drilling, but also reduce our dependence on the corrupt and illiberal governments that often control the supply of these valuable metals.

It is difficult to harvest these metals. Extracting them from the 'unholy sludge', as Blundy memorably put it, requires cutting-edge chemistry and a lot of trade secrets. Turning his monitor, he showed me a picture of a pipe so scaly that you'd think it came from Triton's lavatory. The more mineral-rich geothermal fluids are highly corrosive, Blundy said, and liable to scale the wells' interior. Understandably, geothermal engineers prefer having less salty fluid in their pipes, so Blundy is working on designing novel materials that prevent the formation of scale. He is also working on existing or prospective mine-the-brine projects in places such as Japan and Montserrat. Blundy explained why we get mineable clusters of individual ores, a story that starts with the elements' creation in exploding stars and continues with the slow but ineluctable activity going on within Earth even today. Different minerals respond differently to heat, pressure and their chemical environment, melting or dissolving in some places and floating around, only to precipitate somewhere else. For 4.6 billion years, Blundy said, Earth has been like 'some crazy librarian, slowly but surely putting all the lithium here, and all the mercury there, and so on'. It is to this process that we owe everything we've ever mined.

The subterranean sorting process also explains why there is, according to Blundy's estimate, enough valuable metal beneath every volcano to make these reserves comparable in value to major conventional mines. The idea wouldn't apply to places like Cornwall, and it wouldn't be financially viable on its own.

But brine-mining is a plausible way of making conventional geothermal, which is expensive to get going, a little more financially attractive.

Crack the nut of geothermal power and it will feed us for billions of years: the Sun will engulf us long before Earth's core stops providing us with heat. In the here and now, a successful geothermal industry would mean a neat repurposing of oil and gas infrastructure and expertise; little prospect of Putin-style energy blackmail; and, most importantly, abundant clean energy, available 24/7, regardless of geography. Perhaps equally thrillingly, we would have drilling that would make the Soviets' Arctic Circle record-breaker look like a hobbit hole.

CHAPTER 8
THE MISFIT PRODIGY

Thought bubbles
Inflating multiplying drifting
Rearranging and annexing
Little rainbows of wishful thinking
Enclosing gases of concern

Nick Bostrom, 'Bubble Ontology'

On 16 April 2024, the website of the Future of Humanity Institute was replaced by a simple landing page and a four-paragraph statement.[1] The institute had closed down, the statement explained, after 19 years of existence. The statement briefly sketched the institute's history, appraised its record, and referred to 'increasing administrative headwinds' blowing from the University of Oxford's Faculty of Philosophy, in which the institute, generally referred to as FHI, had been housed.

Thus died one of the quirkiest and most ambitious academic institutes in the world. FHI's mission had been to study big-picture questions for humanity: our direst perils, our range of potential destinies, our unknown unknowns. By the time it was founded, we knew about the natural risks, climate change and nuclear war. But what else did we need to know? FHI's researchers pondered this matter deeply. The concept of super-intelligent AI is one of several that they ushered into bestseller lists, serious academic study and the headquarters of the United Nations.

To its many fans, the closure of FHI was startling. This group of polymaths and eccentrics, led by the visionary philosopher Nick Bostrom, had seeded entire fields of study, alerted the world to grave dangers, and made academia's boldest attempts to see into the far, far future. As well as fans, though, the institute had detractors. The one thing FHI had not foreseen, those detractors commented archly, was its own demise.

Why would the university shutter such an influential institute? Was this a whodunnit? What was FHI like on the inside? To what extent did its office harbour medieval weaponry? And, to invert the organisation's mission somewhat, how, if at all, will our descendants look back on FHI? For an institute that set out to find answers, FHI left observers with a lot of questions.

In 1989, a teenage Swede borrowed a library book of 19th-century German philosophy, took it to a favourite forest clearing and experienced what *The New Yorker* would later describe as 'a euphoric insight into the possibilities of learning and achievement'.[2] Damascene moments are rare in real life, but this seems to have been one of them. Niklas Boström dedicated himself to a lifetime of intensive study. He withdrew himself from school, rattling through syllabuses at home, and he read widely and manically. At the University of Gothenburg, he took a BA in philosophy, mathematics, mathematical logic and artificial intelligence.[3] After that, he took postgraduate degrees in philosophy and in physics, in astrophysics and general relativity, and in computational neuroscience. In what little spare time he had, Boström emailed, and met up with, fellow transhumanists: an assembly of vitamin-popping free spirits, united by their enthusiasm for radically improving human biology and lifespans via methods such as cryopreservation.

In 1998, Boström co-founded the World Transhumanist

Association.[4]* As early as 2001, he was studying little-known phenomena called 'existential risks', writing that nanotechnology and machine intelligence could one day interfere with our species' ascent to a transhuman future.[5] He also, in 2003, formulated the 'simulation hypothesis', advancing in *Philosophical Quarterly* the theory that we might be living in a computer simulation run by humanity's hyper-intelligent descendants.[6]† Most famously, he posited the thought experiment of the 'paperclip maximiser', illustrating the risk posed by powerful AI by imagining such a system being tasked with improving the output of a paperclip factory. If the system is able to improve itself in service of its goal, wrote Bostrom, what follows is 'the consequence that it starts transforming first all of earth and then increasing portions of space into paperclip manufacturing facilities'.[8]

By 2003, Bostrom‡ had arrived at Oxford as a postdoctoral fellow at the Faculty of Philosophy. Many years later, the faculty would be his *bête noire*. But it was Bostrom's membership of it that enabled a stroke of luck that would dramatically change his life. He had already come across James Martin, an IT entrepreneur. Following his success in the business world, Martin had become a prescient futurist. He wrote more than 100 books, including a Pulitzer-nominated exploration of how computerisation would change the world.[9] Martin also produced a documentary featuring Bostrom.

But Martin, who lived on his own private Bermudan island,

* The transhumanists deserve a much fuller treatment than I am able to give them. I recommend that those interested read Elise Bohan's PhD thesis, which is a rollicking history of this group.[7]

† Armchair philosophers will identify that this theory is itself a descendant of similar thought experiments. Plato proposed that our relation to reality might be that of cave-dwellers to the outside world. Descartes posited, in place of hyper-intelligent descendants, an evil demon. And we must not ignore the contribution made to this field by *The Matrix* (1999).

‡ Time and Anglicisation have sanded off his umlaut.

was not only a producer; he was also a deep-pocketed philanthropist. Thanks to Julian Savulescu, who was another young philosopher interested in human enhancement, Bostrom learned that Martin was planning to fund future-minded research at Oxford. Hoping that this could encompass work on his specialisms, Bostrom made his case to the university's development office.

After two decades, the details are hazy. Bostrom emphasises that there was no single turning point. But FHI lore has it that the university liked to host dinners where such donors sat next to exciting academics, creating the perfect conditions for what we now call a nerdsnipe. Bostrom spoke earnestly and intelligently about existential risk and transhumanism, goes the story;* Martin was deeply engaged. 'That's Nick's greatest charm, actually,' Bostrom's FHI colleague Anders Sandberg told me. 'He can be nice, but he's even better when he's just going at an important question, explaining why it is the most important thing in the universe.' Martin was, according to an FHI old-stager, 'blown away'.

In 2005, Martin made what was then the biggest benefaction to the University of Oxford in its nine-century history. It came to more than £70 million,[10] of which a small portion funded what Bostrom termed the Future of Humanity Institute. 'It was a little space,' Bostrom told me of this nascent FHI, 'where one could focus full-time on these big-picture questions.'

That seed grant was small: enough for a few people for three years. Narrow thinkers were not welcome. With such a small team, and such an unconventional brief, Bostrom wanted multidisciplinarians. He was looking, he told me, for 'brainpower especially, and then also a willingness and ability to work in areas where there is not yet a very clear methodology or a clear

* Bostrom can no longer recall which dinner happened when, and Martin is no longer with us. This leaves us with only the second-hand anecdote to rely on.

paradigm. So, it's a generalised thinking skill, often combined with some kind of polymathy.'

One of his earliest hires was Anders Sandberg, whom you might recall from my chapter on super-eruptions. Sandberg, also a Swede, was a fellow member of the Extropians, an online transhumanist community that Bostrom had joined in the 1990s.* This community also included Eliezer Yudkowsky, a brilliant, disagreeable young autodidact† who often corresponded with Bostrom, as well as individuals who helped initiate the cryptocurrency revolution.

Where Bostrom is generally ultra-serious, Sandberg is ebullient and whimsical. These differences in personality belied the two men's similarity in outlook. Sandberg, too, was an unorthodox thinker who was interested in transhumanism and artificial intelligence. (Within transhumanism, Sandberg was particularly interested in the theoretical practice of whole-brain emulation; that is, the uploading of a human mind to a digital substrate. This might, one day, be humanity's best long-term solution to biological death.)

After finishing a PhD in computational neuroscience at the University of Stockholm, Sandberg had embarked on an exhibition project in which he tried to explain the human brain 'to everyone,' he told me, 'from kindergarten kids to professors'.

* The name 'Extropian' is a play on the word 'entropy'. Entropy, put simply, is the natural tendency toward decay. An Extropian believes that living beings ought to be able to defy entropy and remain indefinitely vital.

† It would be impossible to do justice in this footnote to the singular mind that is Yudkowsky's. I will return to him later, but for now, know this. As a teenager, he quit high school in order to teach himself, scoring preposterously highly on his exams. As an adult, his works on rationality and AI have made him one of the most influential thinkers of our time. He is a lordly essayist, sometimes given to administering brutal put-downs to those in his intellectual warpath. As he told one such victim in 2023: 'You are trying to solve the wrong problem using the wrong methods based on a wrong model of the world derived from poor thinking and unfortunately all of your mistakes have failed to cancel out.'

With that project finished too, Sandberg was free to be interviewed at Oxford, arriving on a wet day in July 2005. Even a futurist, it turns out, can appreciate the timelessness of the dreaming spires. 'I was walking along Queen's Lane in the rain, and I realised that I couldn't tell, from this vantage point, which century I was in.'

Sandberg was interviewed in the Faculty of Philosophy's Ryle room, named for the philosopher Gilbert Ryle.* It is a room, Sandberg explained, where philosophers would have smoked pipes for decades. 'Smoking had probably been banned there for decades too, but the philosophical smoke was still in the walls. And it seemed like it worked.' Sandberg explained some neuroscience to the faculty staff who were assessing him and communicated his aptitude in another little-known area of human endeavour: web design. He was hired, and he returned in January 2006 to take up a desk at FHI and a 'silly little room in Derek's house'.

This was no ordinary Derek. Sandberg was lodging, with Bostrom, in the home of Derek Parfit, a wild-haired recluse who was also one of the most influential moral philosophers of the modern era. Bostrom had the master bedroom and collected rent from the rotating cast of lodgers.† Parfit, Sandberg recalled,‡ slept in 'a little cubby hole' of a bedroom and would scuttle at odd hours between the cubby hole and his office at All Souls, the highly selective graduate college seen as an Oxford within Oxford. For a time, the house's proverbial trousers were worn by a mouse that nibbled the philosophers' food and

* Ryle, famously, was critical of the view that the mind is a mysterious and non-physical entity that nevertheless controls the body. Perhaps he might have been sympathetic to the idea of whole-brain emulation.

† Sandberg and Bostrom were part of a succession of young intellectuals who lodged with Parfit. David Edmonds, in his biography of Parfit, describes 'a variant of the hit TV comedy *The Big Bang Theory*, except that, instead of being a household of physicists, it was chock-full of philosophers'.[11]

‡ Parfit died in 2017.

consistently evaded their ethical, but ineffective, attempts to trap it. 'The mouse was having the time of its life,' said Sandberg, 'up until the point where I accidentally killed it when throwing away some garbage. I had to bury it behind the Ashmolean Museum.'

Though the mouse did not live to see FHI take shape, the institute was developing nevertheless. Including Sandberg, Bostrom hired three researchers and began to sculpt a research agenda that, in these early years, was primarily concerned with the ethics of human enhancement. An EU-funded project on cognitive enhancement was one of FHI's main focuses in this period. The institute also organised a workshop that resulted in Sandberg and Bostrom authoring an influential paper that laid out the technological advances required to make whole-brain emulation feasible.[12] At the same time, FHI staff were producing work on the gravest perils facing humanity, a topic that was not yet an established academic discipline. In 2008, Bostrom co-edited the book *Global Catastrophic Risks*, which brought together analyses of such threats as asteroid impacts, nuclear war, totalitarianism and advanced nanotechnology. Bostrom would soon co-author research with Eliezer Yudkowsky, the brilliant Extropian whose excitement at the possibilities of AI had turned to terror.[13] Yudkowsky had hitherto existed outside the academic firmament; his co-authorship with Bostrom symbolised the entrance of his unorthodox ideas into the august halls of academia.

Those who worked for FHI had to accept temporary contracts and an ugly office building, Littlegate House. The building bore scant comparison with the beautiful quadrangles that were its near-neighbours. (Jaan Tallinn, an occasional visitor, is said to have joked that the windowlessness of Littlegate House had a hidden benefit – it would make FHI easier to simulate on the graphics cards of humanity's descendants.) In return for his staff's forbearance, Bostrom tried to lay over them a carapace that would shield them from many of the more tiresome demands of academic life. There were no requirements to teach. As time went by, there was decreasing pressure to publish via traditional academic modes. Explaining this philosophy to me, Bostrom compared his staff to gems so brilliant that a jeweller would want to create bespoke casting for them. He wanted 'to pick these jewels and then create a kind of organisational fixture around them that would let them scintillate and do their thing with the smallest possible number of distractions'.

One of Bostrom's first selections was Toby Ord, a computer scientist turned philosopher. Ord was one of the key figures in the founding of Effective Altruism. His life had changed when he read the 'drowning child' argument advanced by the Australian philosopher Peter Singer. If you'd be happy to ruin a coat by leaping into a pond to save a drowning child, goes the argument – and many readers might nod their heads at this point – you should also be happy to spend the coat's worth in order to save the life of a child far away from you. At this point, readers often stop nodding; Ord did not. He researched the most cost-effective charities in the world, finding them to be a selection of those that addressed neglected diseases in the developing world. And he pledged to give away everything he earned over a modest inflation-adjusted allowance.

In 2009, Ord was put in touch with a similarly-minded graduate student called Will Crouch.* A coffee meeting in a graveyard, of all places, became a five-hour discussion of the necessity of giving, and the best ways in which to do it.[14] Crouch helped Ord found a charity, Giving What We Can, whose members give 10 per cent of their earnings to charities deemed to be effective. This meeting was a crucial moment in the development of effective altruism – a movement that would later provide considerable financial muscle to the thinking that emerged from FHI.

* Having changed his name on getting married in 2013, Will Crouch is now known as Will MacAskill.

Another early hire was Eric Drexler, an engineer, Extropian and futurist who has been called 'the undisputed godfather of nanotechnology'.[15] Like Ord and Sandberg, Drexler would stay with FHI until the very end. Each of the senior researchers had their own signature style of thinking. Bostrom would turn roiling, nebulous concepts into categories, like a cartographer mapping untouched continents. Ord would relentlessly quantify, often grappling with different forms of infinity. Drexler would approach things as an engineer, starting with a blueprint before moving on to practicalities. Together with Sandberg, they helped set the institute's tone: eclectic, curious and unabashedly vigorous in their pursuit of questions other academics would not touch.

Outside the carapace, the world was changing. It was the early 2010s and 'AI winter', as the field's period of moribundity is known, was easing. FHI staff, Bostrom in particular, began to spend more time considering the risks and opportunities that might result from the era's advances in machine learning. Bostrom began writing a book of his own on catastrophic risks; the work on human enhancement was largely wound down. 'This was a fairly typical approach for FHI,' Sandberg wrote in his retrospective of FHI.[16] The *modus operandi* was to find a neglected topic deserving of research, before 'germinating it in the sheltered FHI greenhouse; showing that progress could be made; coalescing a field and setting research directions; attracting bright minds to it; and once it's established enough, setting it free, and moving onto the next seedlings'.

Two particular seedlings – AI risk and AI governance – were to become almighty forests. FHI itself was now a quickly growing sapling. In 2011 or so, it had been moved downstairs to a bigger set of rooms within Littlegate House. The original premises had whiteboards, but Sandberg had insisted on more. The new office had a central room that was encircled by whiteboards – a 'breathtaking 200-degree panorama!', as Sandberg put it.[17] It was nicknamed the 'whiteboard panopticon'. This room, the James Martin Room, became the epicentre of the institute's collective thinking. Here, FHI staff would sketch out anything from potential solutions to AI safety (one researcher was briefly overjoyed when he mistakenly believed he had solved the problem), to a prediction of what interstellar war would involve, to a consideration of what music would be like if we had more than one time dimension. (The latter exploration was, of course, Sandberg's. Operations staff would sometimes groan at researchers' more flagrant indulgences of whimsy.) Particularly interesting whiteboard notes, especially those which involved collaboration, would be left up for days or weeks. On another whiteboard, Bostrom maintained a scorecard for FHI. He worked at night and was believed to update the number when nobody else was around. Over time, the number crept upward, though it would fall when there were setbacks. Bostrom, who can call upon a dry wit when he chooses to,* told me that the precise workings of the metric were 'shrouded in mystery'.

Martin died in 2013. FHI commemorated its most important early supporter by hanging a picture of him in the room that bore his name. Martin's death, to the sorrow of FHI staff, meant that he missed what was to be FHI's heyday. The institute was becoming rapidly more influential, and its staff's mistakes – historic and ongoing – were yet to catch up with them.

A crucial factor in FHI's ascent to prominence was its earlier decision to focus more intellectual energy on the risk posed by AI. Bostrom, you will recall, had been writing a book on catastrophic risks. A chapter of that book had taken on a mind of its own, so to speak, becoming the sole focus of the project. The chapter, naturally, was on artificial intelligence, and the book, the bestselling *Superintelligence*, warned humanity that the

* Internal FHI legend had it that Bostrom, as a younger man, did stand-up comedy. It was even claimed that there was extant footage, though few staff, if any, ever saw it.

creation of super-intelligent AI is likely to be our final act – for good or ill.

Bostrom likes to adorn his work with fables of his own devising and the book's cover, a beady-eyed owl, refers to a story he included in the book. A group of sparrows, wishing to build their nests with the help of a bigger bird, set out to find an owl egg. In doing so, they defy the warnings of Scronkfinkle, 'a one-eyed sparrow with a fretful temperament'. Scronkfinkle tells his fellow sparrows that, without an attempt to master owl-taming, the plan would be their ruin. The sparrows do not listen and the fable is marked unfinished.[18]

Thanks in part to Bostrom's knack for storytelling, the book was a wild success. Elon Musk, who would soon become a donor to FHI, publicly praised it, as did Bill Gates. (Musk's donation was the first major funding for technical work on AI safety, though his enthusiasm would later be used against FHI.[19]) Derek Parfit was said to have received the book as a 'work of importance', while Sam Altman, then a 29-year-old who had just been put in charge of the Silicon Valley start-up accelerator Y-Combinator, wrote that *Superintelligence* was 'the best thing I've seen' on the topic of AI risk.[20] In October 2015, Bostrom briefed a United Nations committee on the dangers posed by future technologies.[21]

When academics become famous, they can sometimes be regarded with suspicion by colleagues. But this was not true of Bostrom at FHI, where he was deeply respected. 'From the outside,' said a former colleague, 'I wouldn't have been able to see the difference between Nick and the other researchers. It's only when you watch them in discussion that you see it. Oh my God, the long tail of intelligence really is long.'

Littlegate House, that cramped, ugly building with a dearth of windows, now felt like one of the brightest intellectual scenes in the world. The work on AI was more than talk: FHI researchers were among the first people in the world to do empirical, rather than just theoretical, work on the problem of AI alignment. After a period at FHI, Jan Leike and Paul Christiano helped create the method we now know as reinforcement learning from human feedback, a method which today undergirds every major large language model. With two research scholars, the alignment specialist Owain Evans provided an important benchmark of AI truthfulness; this benchmark is still used by major developers.[22] And Katja Grace, with Evans and others, began the project that became AI Impacts.[23] (AI Impacts gathers and synthesises experts' views on what we can expect from AI development, providing useful data for decision-makers.)

By the mid-2010s, FHI was attracting visits not only from technology heavyweights such as Demis Hassibis (co-founder of DeepMind) and Vitalik Buterin (inventor of Ether), but also curious hacks from the mainstream media. The writer of Bostrom's *New Yorker* profile found the FHI office to be 'part physics lab, part college dorm room', noting the posters of the film *Brave New World* and of HAL 9000, the computer that goes rogue in *2001: A Space Odyssey*.

Visiting in the same year, a breathless young newspaper reporter* described FHI as being akin to 'the School of Athens taking place in an IT support office'. There were split keyboards, homemade keyboards, Dvorak keyboards.† Embedded in furniture were loose Nerf gun pellets, the remnant of a day in which the young daughter of Stuart Armstrong, an AI safety specialist, had gone hunting for her father's fellow researchers. ('The way of getting a little girl out of her shell is to give her a gun,' Sandberg observed to me, 'or at least a Nerf gun.') Two of the rooms bore names that might be familiar. One was named after Stanislav Petrov, the Soviet lieutenant colonel who averted a nuclear war by correctly guessing that what appeared to be a hail of incoming American missiles was in fact a computer

* Me.

† Dvorak keyboards, unlike Qwerty keyboards, have their keys arranged for optimal typing speed.

glitch. Another was named after Vasili Arkhipov, the submariner who vetoed the use of a nuclear torpedo in the Cuban Missile Crisis.*

Ord's office was beside a thoroughfare. When I visited, I knocked on his door and he gamely chatted to me about the universe's deep future, in which our descendants might harness entire galaxies to power our computation. Sandberg's was a grotto of curios. He had a collection of chemical elements, a *Terminator* skull and many other trinkets. (Bostrom joked that he had saved Sandberg from becoming a museum curator.) Having heard a rumour that there had once been a sword in Littlegate House, I asked Sandberg if it was his. It wasn't, he said, but the rumour might have had something to it. Stuart Armstrong had a collection of katana swords at home and Bostrom, in the days of the Parfit house share, dabbled in swordsmanship himself. On hearing a ruckus from the street one night, Sandberg looked out of his window to see Bostrom and his future wife, Susan, duelling with foam swords they had won at St Giles Fair.

At FHI, Bostrom's ludic side was less visible. In the tiny kitchen – which wasn't really big enough for more than six – more than a dozen researchers, typically, would convivially cram in for vegan lunches brought in from a nearby supermarket. Bostrom would come in later, having designed his nocturnal working pattern so that he could better stay in touch with Susan, who lived and worked in Montreal, and their young son. (Bostrom spends six months per year with his wife and child. For the other six months, the nocturnal working pattern has the additional advantage of quietness and solitude.) Arriving at FHI in the afternoon, he would work into the small hours. But first, in the afternoon, he could be spotted in the kitchen. Here he would concoct the vegetable-based smoothie that he wryly

* I told these stories in more detail in Chapter 4.

called his 'elixir'.* This was often the only time that staff would bother him before he disappeared into his office. Knowing that Bostrom liked to descend deep into the halls of concentration, staff would seldom disturb him. Tanya Singh, whose five-year stint at FHI encompassed several senior operations roles, as well as periods of being Bostrom's executive assistant, said she knocked on his door only seven or eight times in all that time.

On the rare occasions that she did enter Bostrom's brightly lit room,† Singh would find him sitting and thinking in near-perfect stillness. 'There was a palpable intensity in that stillness,' she said. 'I have never seen anything like it. You could drop something next to him – a bomb could go off – and he wouldn't move, he wouldn't register it at all.' (Another member of FHI staff recalled entering the whiteboard room with a group to find Bostrom sitting deep in thought. The group quietly left the room.) Interruptions, Singh believed, would ruin Bostrom's day 'because he would have to yank himself out of that mind state'. Bostrom was as protective of his own freedom to sit and think undisturbed as he was of his staff's. It was a carapace within a carapace, and Bostrom, by all accounts, spent little time maintaining relations with the philosophy faculty.

For a while, this didn't seem to matter. FHI's work was becoming even more relevant to the outside world, even if it wasn't much appreciated within Oxford. In March 2020, Ord had published *The Precipice*, a book that examined the perils facing humanity. 'As the gap between our power and our wisdom grows,' he wrote, 'our future is subject to an ever-increasing level of risk.' The book examined perils such as man-made pandemics, to which FHI had begun to devote more time and attention. FHI brought together policy leaders and experts to discuss such threats, and was beginning to map this new landscape of biological risk. Soon after FHI's biological

* Known ingredients: cabbage, cauliflower, carrot, lime and olive oil.

† A *New Yorker*-certified 14 lamps.

threats team had begun drawing attention to an epidemic in Wuhan, much of the world was plunged into the first COVID lockdowns. It was a vindication of a sort, if a grim one. As with artificial intelligence, FHI had been early to sound the alarm.

Philanthropic funders admired this and other lines of work on catastrophic risks. Thanks to these funders' munificence, the institute was expanding. Its administrative duties were taken on by bright young minds who could otherwise have been earning vast corporate salaries or taking on high-status research work at other institutes. As Bostrom put it: 'We had some guy who came with a jurisprudence doctorate from Yale Law School, which is roughly the best law school in the world, initially to work as an unpaid intern to do menial office chores – replace the paper in their coffee machine and stuff like that. He eventually became like a low-level administrator, a job that would pay next to nothing, because of university salary rules. He could have just walked into the most prestigious law firms in Manhattan.'

And recruits kept coming. FHI's Research Scholars Programme, led by the mathematician Owen Cotton-Barratt, was a revolving door of young talent; Cotton-Barratt was renowned at FHI for his ability to spot people with potential and to encourage them into productive lines of research. Other entry points were the DPhil Scholarship and the summer programme for undergraduates. 'Desks were crammed into every conceivable space with increasing ingenuity,' wrote Sandberg. Sometimes Sandberg struggled to make it to his office without being attracted to an intellectual discussion. If novel ideas and conversations are pollen, then Sandberg was a bumble bee, dragged this way and that by colourful stimuli. 'Quite often,' he told me, 'I stood around with my raincoat still dripping as someone was enthusiastically explaining something.' Multiple staff were invited to present their work to the British parliament; Toby Ord would soon to be quoted in the address that Boris Johnson, then the British prime minister, made in 2021 to the UN General Assembly.

But the writing, appropriately for FHI, was already on the wall. In late 2020, and to the shock of FHI, the university imposed on it a hiring freeze. No new staff, no new research scholars. This was one of several measures, including a fundraising freeze, that FHI staff believed were designed to throttle their work. If this was bureaucratic animus, as FHI believed it to be, where did it come from? FHI staff, for their part, often expressed frustration at the delays and paperwork that university membership entailed. As Sandberg wrote: 'One of our administrators developed a joke measurement unit, "the Oxford". 1 Oxford is the amount of work it takes to read and write 308 emails. This is the actual administrative effort it took for FHI to have a small grant disbursed into its account within the Philosophy Faculty so that we could start using it – after both the funder and the University had already approved the grant.'

Few people relish paperwork, but to those convinced of the merits of Bostrom's 'astronomical waste' argument, inefficiency is a profound wrong. (In the paper of the same name, Bostrom argues that the future could contain such profound amounts of happiness that any delay constitutes a loss of value – wastage on an astronomical scale, in other words – that 'boggles the mind'.*) Bostrom wanted to hire people quickly, work with businesses and non-profits, and host conferences without having

* Bostrom's introduction to this paper is the most arresting you'll read for some time. 'As I write these words, suns are illuminating and heating empty rooms, unused energy is being flushed down black holes, and our great common endowment of negentropy is being irreversibly degraded into entropy on a cosmic scale. These are resources that an advanced civilization could have used to create value-structures, such as sentient beings living worthwhile lives. The rate of this loss,' Bostrom continues, 'boggles the mind. One recent paper speculates, using loose theoretical considerations based on the rate of increase of entropy, that the loss of potential human lives in our own galactic supercluster is at least $\sim 10^{46}$ per century of delayed colonization.' As he notes in *Deep Utopia*, we lose three galaxies per year to the expansion of the universe. Send a signal to those galaxies and it will never reach them. We are sundered forever.

to parlay any of this through the university's bureaucratic machinery. But the Faculty of Philosophy, by Bostrom's description, had 'a very different cultural mindset'. Its attitude to hiring, Bostrom told me, was rooted in a culture of teaching the same sort of philosophy – Aristotle, Plato et al – for centuries. '"We have this person who should teach ancient philosophy,"' he said, approximating the faculty view, '"and then when they retire, 40 years from now, we'll hire another person to teach ancient philosophy" ... whereas our research agenda was very much designed to be flexible.' On one occasion, a former member of Bostrom's team told me, the faculty's slow movement cost the institute a new hire. Another alumnus wondered whether, in an academic world in which status is often zero-sum, FHI became a target of jealousy or resentment.

FHI might have been bureaucratically sinned against, but it was also a bureaucratic sinner. It seems to have had a reputation of being difficult to deal with, and – depending on whom you ask – of having management that thought itself above the petty demands of university bureaucracy. 'During my time,' a former senior staff member, Seán Ó hÉigeartaigh, wrote on the Effective Altruism forum, 'FHI constantly incurred heavy costs for being uncooperative.'[24] Its misdeeds, though minor, irritated the university. Staff were reproved for using Gmail instead of Outlook, for travelling without risk assessments, and so on.

The pairing of the institutions was an increasingly unwieldy one. FHI had become larger than the body that housed it, attracted more attention and funding, and employed many more non-philosophers than it did philosophers. Nor, it seems, was the faculty particularly enamoured of the institute. 'The impression I got,' I was told by a don from a different department, 'was that the philosophers' – that is, those within the faculty – 'didn't have much regard for it.'

Fundamentally, it seems, there was a mismatch between the way the organisations assessed the value of their work. 'The philosophy faculty's currency is peer-reviewed papers in prestigious journals that get cited a lot,' said Niel Bowerman, who was assistant director of FHI at the time I visited, 'whereas that wasn't the currency of FHI. The currency was cool ideas that could improve the world.' The mismatch only increased as FHI became more high-profile and started to attract funding with fewer obligations.*

Multiple FHI staff told me that relations had worsened when a new faculty chair, Chris Timpson, arrived in 2018. Timpson chose not to comment when I attempted to speak to him. When I put a detailed set of questions to the university, I was issued with the same vague statement it had released on FHI's closure.† The eternal bureaucratic logjam was, in the view of FHI staff, having real consequences: not just in the more faraway astronomical waste sense, but in the sense of costing present-day lives. Jan Kulveit, who worked at FHI between 2018 and 2023, had led a successful COVID-19 forecasting project in the virus' first wave. He wanted to expand the project and provide medium-range forecasts for the whole world, warning about the potential second wave. The project was offered philanthropic funding, but it turned out it would be too bureaucratically difficult for the university to accept the grant, not least because it would have been used to hire external software engineers. The project expansion didn't happen. 'This was quite frustrating,' Kulveit told me. 'We had the best forecasters and some idea about what was coming. Overall, I think there was some

* In the early days, FHI was obliged to carry out insurance-related research as a result of its partial funding by the re-insurer MS Amlin. I'm told that Sandberg, true to character, managed to find the project ravishingly interesting.

† That vague statement is as follows: 'We regularly consider the best structures for conducting our academic research, as part of the University's governance processes. After such consideration, the decision was made to close the Future of Humanity Institute. The University recognises the Institute's important contribution to this emerging field, which researchers elsewhere across the University are likely to continue.'

mismatch between the timescale of the crisis and timescales on which things like processing grants in academia operate, and I became worried this would be true for some other risks FHI studied.'

FHI asked its biggest donor, Open Philanthropy, to put it to the university's vice-chancellor that the institute have more autonomy.* This, FHI staff told me, did not go down well with Oxford. Within FHI, there was plenty of frustration with the university, but there was frustration with Bostrom, too. In August 2021, Owen Cotton-Barratt, the architect of the Research Scholars Programme, quit FHI. In his resignation letter, addressed to Bostrom and shared with FHI staff and some allies of the institute, Cotton-Barratt praised Bostrom's intellectual leadership, but criticised his management.

On my request, Cotton-Barratt showed me the letter. Its tone was gentle and conciliatory, but its substance was serious. Bostrom was a bad delegator, Cotton-Barratt wrote, and disinclined to invest in communication with staff. By Cotton-Barratt's account, Bostrom prioritised his own research over FHI's relationship with the university; as a result, the institute had suffered. To the dismay of FHI staff, the letter eventually reached the philosophy faculty. (This was not Cotton-Barratt's doing.) Reflecting on the letter, Cotton-Barratt told me in 2024: 'I think Nick did a great job building FHI and the world has lost somewhere special. I wrote that letter in the hope that it might help FHI to iterate towards the best version of itself.' Bostrom viewed the letter as well-meaning, but 'a bit of a facepalm'. It arose from an effective altruist culture of extreme openness, he said, but it 'set us back administratively by about a year'.

After the worst of the pandemic, FHI staff re-assembled at new premises. The institute's home was now Trajan House, an office

* Open Philanthropy is the effective altruism-inspired philanthropic vehicle of Dustin Moskovitz and his wife Cari Tuna. Moskovitz was one of the co-founders of Facebook.

on the outskirts of Oxford. (It overlooked a graveyard, prompting Bostrom, true to his transhumanist roots, to call the building 'Corpsewatch Manor'.[25]) FHI was to share the building with the Centre for Effective Altruism (CEA) and similarly minded non-profits. There were Huel shakes in the fridge* and a nap room that university employees, such as FHI staff, were banned, for bureaucratic reasons, from using. A former CEA employee told me that the university and the Faculty of Philosophy had rarely shown much interest, or pride, in Effective Altruism, despite it being 'one of the biggest success stories in applied philosophy in maybe 100 years'. In the view of this employee, the university had 'a weird beef that has to be motivated by a personal grudge'. (It's also worth considering that EA's uncompromising utilitarianism has been known to get people's backs up.)

FHI's relocation to EA HQ, therefore, might not have been helpful for its relations with the university, but it made intellectual sense. When Toby Ord had helped found what we now know as EA, its concern had been poverty in the developing world. (Fin Moorhouse, who worked alongside him for two years as a young researcher, told me that Ord remains, in moral terms, 'the real deal' – ascetic, kind and intellectually earnest.) As technology such as AI advanced, EA had, like its co-progenitor Ord, embraced the idea that the key moral priority of our time is the protection and improvement of the long-run future.[26] This school of thought – longtermism – is indebted to the work of Derek Parfit, the economist John Broome and Bostrom.† Close to its heart is the endeavour to reduce existential risk. EA, a

* Huel is a meal replacement shake: beige in colour, chalky in constitution and near-impeccable in its environmental credentials. Its name is a portmanteau of 'human fuel'. As a younger man, I was forced by a newspaper to live off Huel for five days. I do not remember that experience particularly fondly.

† In 'The Case for Strong Longtermism', Hilary Greaves and Will MacAskill, two leading EA philosophers, refer specifically to Bostrom's paper 'Astronomical Waste'.

community whose existence was indebted to FHI staff, had, for several years, been turning its considerable resources and talent towards the reduction of existential risk – the overwhelming priority of FHI.

There was no whiteboard panopticon in FHI's new home, and the new office rooms, stacked for months with unopened cardboard boxes, weren't much more spacious than the last ones. Senior staff were seen on-site less often than they used to be, and the hiring freeze – which briefly thawed every now and then on a negotiated basis – still constrained FHI to a painful degree. Tanya Singh, in particular, was working hours that would put a paperclip maximiser to shame. She left in June 2022, the institute's headcount having fallen to the same number as it was in 2017, when she had arrived. But the ascent of Sam Bankman-Fried, an EA crypto mogul who had suddenly become the world's youngest billionaire, must have felt like an exceptionally promising development.

For a single, delirious summer, it seemed that every other person at Trajan House, including FHI staff, was to-ing and fro-ing between Oxford and the Bahamas. There, Bankman-Fried and his allies were planning a barrage of longtermist philanthropy. 'The mood was pretty buoyant,' recalled Lewis Hammond, an AI researcher who had joined FHI in 2019. Ord, apparently, was one of the few who evinced reluctance to get involved. In October 2022, however, Bankman-Fried's empire disintegrated. Millions of people had been swindled and Bankman-Fried's deceit had tarred his allies: EA, FHI and the many individuals associated with them. One can imagine the disgruntlement in the Ryle room. By now, FHI's seminars and salons were tailing off. When the Stakhanovite Singh departed, the faculty had seconded a part-time administrator who covered a fraction of her hours. 'It felt like FHI was dying a slow death,' said Hammond. Or, as Singh had it: 'Death by a thousand paper cuts.'

That death became a painful one. On 9 January 2023, Sandberg posted to Twitter a document written by Bostrom:

'Apology for an Old Email'. The email was sent in 1996 to the Extropians listserv. In a conversation about offensive communication styles, a 23-year-old Bostrom wrote that he had always liked 'the uncompromisingly objective way of speaking'. The more counterintuitive and repugnant a formulation, he wrote, the more it appealed to him, assuming it was correct.

As an example, he wrote that 'Blacks are more stupid than whites', commenting that 'I like that sentence and think it is true'. This did not mean, Bostrom remarked, that he disliked black people and thought it was right that they be treated badly. 'It's just that based on what I have read, I think it is probable that black people have a lower average IQ than mankind in general.' Yet the sentence he'd posited, he wrote, the claim that black people are stupider than whites, would be taken by most people as 'I hate those bloody niggers!!!!'. By Bostrom's account, he apologised for the email within 24 hours of sending it.

Seventeen years later, a longtermist-turned-critic, Émile P. Torres, found the email. Bostrom, informed that Torres was likely to publish the email, published it himself, with an apology, and disseminated the document via Sandberg (Bostrom does not use X/Twitter). 'I completely repudiate this disgusting email,' Bostrom wrote. Explaining his views, he said that 'it is deeply unfair that unequal access to education, nutrients, and basic healthcare leads to inequality in social outcomes, including sometimes disparities in skills and cognitive capacity'. Whether there are genetic or epigenetic contributors to differences between groups in cognitive abilities, he wrote, was not his area of expertise. Torres, and other critics, took the email and apology as evidence for their case that longtermism was motivated by a hateful form of eugenics.

The university suspended Bostrom while it and the faculty investigated him. Some FHI staff, along with outsiders, disapproved of the apology. It was 'insensitive', an FHI alumnus told me, and the controversy as a whole – both the original email and the debate over the apology – had further damaged FHI's reputation. Bostrom, meanwhile, was forbidden to contact his colleagues, or undertake anything FHI-related, for the duration of his suspension. By now, the existence of FHI was plainly fragile. Bostrom's staff wanted clarity from the university, but the university, I was told, refused to speak to them. This is because it was a matter for the director – whom the university had exiled. 'And contrary to popular conception,' says a former member of FHI staff, 'Nick is an extreme stickler for rules.'

The outcome of the investigation, though, was in Bostrom's favour. 'We do not consider you to be a racist or that you hold racist views,' a representative of the university told Bostrom in August 2023, 'and we consider that the apology you posted in January 2023 was sincere.' He returned to official duties.

Throughout this difficult later era for FHI, senior staff hunted for a solution. They discussed the idea of Bostrom retaining directorship of the institute's research, but handing over management to a CEO. They mooted becoming part of one of the university's constituent colleges, or spinning out of the university altogether. They were even in discussions with the Faculty of Physics over an interdepartmental transfer. Yet none of these plans became reality. Staff valued their association with Oxford, making them reluctant to leave the university. As for the interdepartmental transfer, FHI staff suspected that the plan was sabotaged by the Faculty of Philosophy.

'We were in this weird limbo for ages,' Lewis Hammond said. But one day in March 2024, while in Trajan House, Hammond was interrupted at his desk by a university IT worker. 'FHI is closing down,' he was told. 'You're going to need to give that computer back.' The university was finally swinging the axe. On 16 April 2024, FHI was shuttered.*

* I noted at the beginning of the chapter that FHI's detractors had mocked the organisation's lack of foresight into its own demise. Yet the institute had, in fact, foreseen its doom. An FHI veteran disclosed that he had, in 2019, been party to a brainstorming session regarding existential risks to FHI as

The reason given by Oxford, Bostrom reported, was that the university did not have the operational bandwidth to manage FHI. He told me that he asked whether the email incident had anything to do with the decision and that the university said it had not. Oxford has not contradicted this account.

'I wasn't even surprised that I was hearing it from the IT guy,' said Hammond. 'It felt very emblematic of how things were being communicated and organised at that point.' He and his colleagues left the building, and the university staff forbade the other residents of Trajan House from using the empty room. Bostrom, meanwhile, was abroad, working on a new book in an Alpine chalet. FHI's final demise, he later told me, came as a relief. 'I'd imagined the faculty would, as it were, let us bleed out,' he said. After receiving the news, Bostrom returned to work.

Back in Oxford, Toby Ord organised a pub trip that staff now refer to as FHI's 'wake'. The remaining staff, and some friends of the institute, met at the Holly Bush to mark the end of FHI's 19-year-life. Ord delivered the eulogy. FHI's death had been 'over-determined', he said, but it had far outlasted the three-year period for which it had been initially funded. The institute had achieved a lot, and its offspring were alive and well. Maybe there wouldn't be anything that would replace the particular kind of organisation that FHI was, Ord concluded, but maybe that was okay. After the wake, he updated FHI's website with the statement announcing its closure.

* * *

an organisation. 'Basically, it was a big Google doc in which everyone dumped ideas, which could then be up- or down-voted. Many people suggested some version of "administrative headwinds" (including more specific things like the hiring freeze). Amusingly, someone also raised the risk that "too much of the research ends up on private Google docs and so doesn't have much wider impact". This received 10 upvotes (the most any single suggestion received was 15) and appears to have been the fate of the document in question.'

After a funeral come the obituaries. Sandberg wrote FHI's official retrospective;[27] Moorhouse wrote a valediction of his own.[28] Fans turned one of Bostrom's poems[29], one that portrayed super-heroic FHI staff blending in among tweedy dons, into a rousing club anthem.[30] (FHI was a 'misfit prodigy', Bostrom wrote, contending with 'thousand meetings, thousand emails, thousand rules'.) The *Guardian*, making the most of Musk's single donation to FHI, took the opportunity to invoke its favourite folk devil – 'Oxford shuts down institute run by Elon Musk-backed philosopher,' went the headline – and retold the stories of EA's various scandals.[31]

As Ord had pointed out at the wake, FHI was survived by many offspring. Who are they? To the extent that FHI is their parent – it might otherwise be called their midwife – there are rationalism, effective altruism and longtermism. The latter two movements, which overlap enormously, have been responsible for a colossal amount of work and funding being directed towards pandemic prevention, technical research into AI safety and other FHI-style endeavours. More specifically, there are the projects and organisations, such as the Centre for the Governance of AI (GovAI), that span directly out of FHI. GovAI helps shape the way society is managing the development of artificial intelligence, as do the dozens of FHI alumni who now work at the leading commercial labs, in think tanks and in related government agencies. Similarly, what was once an apparently wacky FHI idea – the theory that we should explore the moral patienthood of forthcoming AIs that might be able to feel as well as think – is now being taken more seriously. Bostrom had encouraged his staff to start work on this nascent field.

Had FHI never existed, and therefore failed to lay the early groundwork, the world would still have invented AI governance and would still have got round to considering the possibility of digital sentience. It would be even shorter of talent and expertise, though, and it would have been slower to wake up to the threats of rogue and misused AI. Bostrom and his team are widely and deservedly credited not only with making an effective case for taking AI safety seriously, but for making the study of existential risk a serious academic business. Where FHI was, at the time of its inception, the world's only institute of its kind, it grew to have several cousins. It's worth noting that one of these organisations has survived the Faculty of Philosophy's bureaucracy – namely the Global Priorities Institute. GPI, as it's known, is another relatively young big-picture sort of institute; it's also the one that employs Will MacAskill, the philosopher who is seen, with Ord, as the co-founder of the Effective Altruism movement. GPI is less zany than FHI, and more academically legible.

An academic with links to the Faculty of Philosophy contrasted the fate of FHI with the continued existence of other such institutes. The academic told me: 'I'm not sure if we should find it amusing or terrifying, regarding humanity's prospects of surviving global threats, that this research SAS team was killed by paperwork.' There has also been a proposal to found an 'FHI of the West' – that is, the Western hemisphere – in Berkeley, California.[32] Sandberg told me that our current culture is too risk-averse and pessimistic, and that we might make better progress if we have more things like FHI in the world, rather than fewer. 'Historically, clusters of intense creative intellectual work have counted for a lot – the cafés of 1920s Vienna, Xerox PARC, Los Alamos, various Enlightenment era salons, those golden decades in Athens, and so on. I'm not claiming FHI was anywhere near that level, but when smart people come together and begin bouncing ideas off each other and have the wherewithal to make things, then breakthroughs and new visions bloom. And we need breakthroughs and new visions.'

As for the existing organisations, none of them can claim to be half as philosophically fecund as FHI was. It started early: in 2006, FHI staff helped birth the Rationalist community by co-founding the blog Overcoming Bias, from which arose the forum LessWrong.[33] This forum would become the battlefield for different schools of AI safety, a community whose story I will tell in Chapter 13. Philosophy around AI, alien life and the far, far future is now replete with FHI ideas and coinages. 'Note how much of the literal terminology was coined on (one imagines) a whiteboard,' wrote Fin Moorhouse, who, apart from assisting Ord, conducted research into space governance and into institutions that might support the interests of future generations.[34] He will look back on his time at FHI 'with fondness', he told me. 'It was a crazy and interesting thing to be part of.' Lewis Hammond spoke of FHI in similar terms. 'Overall,' he said, 'my main feeling looking back on FHI is one of gratitude. It was both exciting and humbling to be surrounded by such thoughtful people and important ideas.'

Appropriately for an institute founded by a transhumanist, FHI is living many afterlives. These successes are part of a rich legacy for FHI, though its counterfactual impact, to use another FHI-ish phrase, is not straightforward to analyse. As Bostrom told me, 'it's like the comment about the French Revolution, right? Was it good or bad? It's too early to tell.' On one hand, FHI has, undeniably, led the way in developing our species' understanding of the risks we face and the destinies that might await us. On the other hand, it's possible that CEOs such as Sam Altman, he of the enthusiastic review of *Superintelligence*, might be less alive to the power of advanced general intelligence (AGI) had Bostrom illustrated it less vividly. One of the most vivid of those illustrations was that of the paperclip maximiser, which epitomised a style of forecast about AI that Seth Baum, of the GCRI think tank, believes to be outmoded. The current direction for AI, Baum tells me, seems to be towards a plurality of systems rather than a singular one that, in the style of the paperclip maximiser, breaks out and takes over the world.

'It feels a little harsh to call that a mistake,' said Baum of the forecast of highly agentic AI. At the point when Bostrom was dragging these ideas into the public sphere, Baum explains, meaningful AI was yet to exist. 'And who knows? Maybe the technology will take a turn in a different direction, and now their ideas will seem correct again.' Baum's view was that FHI's years of academic graft compared favourably to the relative dilettantism of commentators such as Stephen Hawking, Jared Diamond and Steven Pinker; each has written on societal collapse and humanity's destiny without acquiring deep expertise in those topics. 'To FHI's credit, and to Nick Bostrom's credit, they have backed up their public prominence with significant research.'

Not everyone agrees that FHI's work was of good quality. Carla Cremer was once an FHI research scholar and is now a softly spoken apostate. 'Research out of FHI might been impactful,' she told me, 'but I think a good part of it was of low quality and maybe even harmful for understanding and acting on global risk.' Cremer told me that one ought not to think of existential risks as discrete, separating them into buckets such as AI, biological threats, climate change and so on. 'Instead,' she said, 'we should think about a level of X-risk* in the system that is heightened by the various actors, institutions, and dynamics that produce this kind of risk.'

To borrow the terminology of the philosopher Daniel Schmactenberger, these actors, institutions and dynamics together comprise 'production functions'. If we apply this way of thinking to catastrophic risks, we might consider zoonotic pandemics – those that originally arise in the animals, such as the billions imprisoned in factory farms. To contain the damage of such pandemics, the X-risk community champions such innovations as next-generation personal protective equipment. But if we look upstream, we will see that society has made zoonotic pandemics much more frequent and severe. The rewards of factory farming are concentrated, making the practice attractive, but the risks are dispersed across humanity. As a result,

* A common abbreviation for existential risk.

factory farms have caused many outbreaks of disease, but we have ignored these signals.

Responding to that example, Cremer said that such situations are analogous to 'a human brain overriding bodily signals of stress'. In the same way, she said, Bostrom's autocratic leadership overrode the signals that things were amiss at FHI. His stubbornness, she said, 'is a quality. It is a quality that you might want in a philosopher, but you might not want it as the head of an institute that needs to navigate a risky landscape that is deeply political.' (The usual FHI response to this sort of criticism is that looking upstream is valid, but unlikely to yield immediate results, especially when conducted by a small institution.)

Oxford's 'administrative headwinds', having blown down FHI, have also dispersed the band of researchers that made FHI famous. Shattered is the carapace, scattered are the jewels – though they have largely ended up in the same field. Stuart Armstrong is the co-founder, and chief mathematician, of a start-up that is trying to develop fundamentally safe artificial intelligence. Eric Drexler continues to work on AI governance and strategy. Anders Sandberg, at the time of writing, was soon to publish *Grand Futures*, a reputedly gigantic tome in which he will map the physical limits of what advanced civilisations can achieve,* and has joined the newly-founded Mimir Center for Long Term Futures Research. Mimir the Norse god was beheaded in wartime before being relied on (while still beheaded) by Odin for his wise counsel. Mimir the independent research institute is based in Stockholm, and looks like it will afford Sandberg a similarly eclectic research brief to the one in which he revelled at FHI. (I'm reminded of the Galápagos finches, of which each species has developed a beak that works perfectly for the food sources on their respective islands, but not

* If you think this book will be about collecting energy from the sun by encasing it in a Dyson sphere, you are not thinking big enough.

elsewhere. Sandberg, improbably, has managed to change islands without having to change his diet.)

At least some of the old band – including Toby Ord and Fin Moorhouse – have, as of 2025, got back together at a nonprofit called the Forethought Foundation. Forethought is a nonprofit whose research focuses on the world's transition to superintelligent AI systems. Forethought, which also employs Will MacAskill, is based in Oxford, but is not affiliated to the university.

Speaking of which: within four months of closing FHI, Oxford announced the establishment of the Human-Centered AI Lab, promising 'a vibrant community for big-picture thinking about a future of AI that enhances human flourishing.' The description made the lab, which would be housed within the Faculty of Philosophy, sound much like the institute we can assume it was designed to replace, albeit with two small exceptions: an entirely new set of staff, and much more obeisance paid to classical philosophy.

When I spoke to Bostrom in 2024, he was midway through the publicity campaign for his own new book, *Deep Utopia*.[35] In the book, Bostrom considers a world in which the development of super-intelligent AI has gone well. Some observers, he told me, have assumed that the penning of a more optimistic follow-up to *Superintelligence* suggests that Bostrom feels a greater bullishness about humanity's prospect of surviving and thriving. Alas. 'We can see the thing with more clarity now,' said Bostrom, 'but there has been no fundamental shift in my thinking.' When he wrote *Superintelligence*, he said, there seemed an urgent need to explore the risks of advanced AI and to catalyse work that might address those risks. 'There seemed less urgency to develop a very granular picture of what the upside could be. And now it seems like time to maybe fill in that other part of the map a bit more.'

Bostrom was characteristically serious, though he occasionally wove into his speech a droll turn of phrase. He was frank in his assessment of the 'intolerable' situation with the faculty, but gentle in his comments on the individuals who worked for it. He was pleased with FHI's development and dissemination of concepts relating to existential risk, and he said of James Martin, whom he'd sat next to at dinner all those years ago, that 'I think we lived up to the spirit of what he had in mind'. Bostrom also expressed his sense of privilege at having worked in such an intellectually lively environment with such talented people.

It was in this conversation that he made his comment about it being too early to tell, as with the French Revolution, whether FHI was a success. Elaborating, he said: 'It's very difficult to know how, ultimately, this shakes out with reactions and counter-reactions and counter-reactions to the counter-reactions and stuff. But if one thinks that it's better for humanity to have some ability to reflect and think and deliberate about these things, then I think we certainly helped accelerate that.'

In that respect, Bostrom's leadership of FHI indisputably changed the world. How did it change him? I expected Bostrom to view this question as a trivial one, but he found in it a kernel of a philosophical question. 'It would be kind of interesting if you could somehow enter a time machine and meet your earlier self. How would they strike you? Because, in reality, things change gradually.' He picked out his 'intensified sense of humility in relation to the big picture – just how profoundly we are in the dark regarding these things. It's not as if we know nothing – we know a whole bunch of things – but it might well be that the things we don't yet know are such that they would radically change the bottom line, as it were, of the overall picture. We are small, and kind of limited in insight and abilities.' He would have felt the same when he founded FHI, he said. The change is that he is now 'inhabiting that sense more actively, perhaps'.

I wondered what awaited him on the other side of the publicity campaign. 'Well,' he said, beginning to smile, 'I'm now a free man, which I very much enjoy. So I'm not in a hurry to put on the chains of bondage again. I can't imagine going back in time. I feel I've had my fill of academia.'

Don't expect to see Bostrom on the golf course. There is no prospect that he will leave his field of study. But *Deep Utopia* might well be the last of his books; they take years to write and Bostrom views shorter projects as more appropriate for our current point in history. AGI is coming, and time is short.

CHAPTER 9
DISEASE X

'With new diseases on our selves we war,
And with new physic, a worse engine far.'

Michel de Montaigne

In a previous chapter, I described how a frustrated physicist directed an intense beam of energy toward something he shouldn't have, thus galvanising a field of study that was not his own. I begin this chapter in similar fashion. If you are a frustrated physicist in charge of high-powered electromagnetic equipment, please take a moment to consider which obstacles to human progress might be better off vaporised.

For geothermal's Paul Woskov, read pandemic prevention's David Brenner. Addressing an audience in 2017, Brenner began by showing an image of some middle-aged football fans.[1] They were Liverpool supporters, bedecked in T-shirts and scarves, and they were sitting around an outdoor dining table on a street in Istanbul. It must be the morning after Liverpool's victory in the Champions League final in 2005.

Brenner then pointed out a smiling man in a red polo shirt. 'That's me,' he said. 'And next to me, in the red hat, that's my friend Paul Rice.'

Rice was a local politician. He was sitting there, also smiling, wearing a Liverpool bucket hat. And then came the kicker. 'A couple of years after this picture was taken, Paul went into hospital for some minor surgery and he developed a superbug-related infection. And he died.'

Brenner is a physicist who, early in his career, specialised in applying the theory of quantum mechanics to radiation therapy. (Radiation therapy is a type of cancer treatment in which beams of intense energy are used to kill cells that have gone rogue.) He had continued to study radiation, emigrating from the UK to New York, where he took the directorship of the Center for Radiological Research at Columbia University Medical Center. The death of his friend Paul Rice, though, prompted a change in direction. 'I declared a war on superbugs,' he explained.

Superbugs are bacteria that are resistant to antibiotics. Antibiotics, of course, have been a transformative technology for humans, and what's miraculous about them is that they specifically target bacteria. Originally derived from microorganisms such as bacteria or fungi, antibiotics affect bacteria by disrupting key functions within them. These key functions can include their synthesis of protein, which is essential to their growth and function, and their replication of DNA, which is how they reproduce. Some antibiotics interfere with cell membranes, which are designed to keep the innards of a bacterium within the bacterium, and the outside world outside it.

What is it like to be a bacterium in the body of a human who has taken an antibiotic? It is probably a little like being a human navigating the Tube system when, without warning, the platforms suddenly start oozing gas. Seeping through your pores, the gas stops your digestion, ceases the workings of your gonads and turns your skin inside out. And, for some reason, the gas is a poison only to you. The trains continue to run and the rats continue to scurry.

This targeted brutality means that antibiotics save millions of human lives per year. Without them, a scratch could mean death. Even basic surgery would become highly dangerous.

By now, antibiotic resistance needs little introduction. It is the phenomenon by which some forms of bacteria have developed resistance to these remarkable drugs. MRSA* is the most widely known of this group. Some of these bacteria produce enzymes that chemically modify antibiotics; some of them have changed their physiology such that antibiotics no longer effectively bind to them; some of them have souped up their efflux pumps, which are structures that expel toxic substances from a bacterium's interior. This is evolution in action and we are accelerating this process by overusing antibiotics both clinically, where they are overprescribed, and agriculturally, where they are routinely used to make factory-farmed animals grow faster.

In both cases, the people doling out the drugs do not directly experience the diffuse negative effects of their actions. The immediate upside is a rejuvenated patient, or slightly fatter animal; the downside is so small and far away as to hardly exist. (Something similar is true of the genesis of every man-made threat to humanity.) But the consequence of antibiotic resistance is that superbugs stalk hospital wards, people like Paul Rice are dying young and superbugs are heading for an annual death toll of 10 million by the middle of this century.

'I'm a physicist,' said Brenner, 'and so I wondered: *Could we take a physics-based approach – a different approach to this problem?* And in that context, the first thing we know for sure is that we actually know how to kill every kind of microbe, every kind of virus, every kind of bacteria. And that's with ultraviolet light.'

Ultraviolet light blasts bacteria and viruses with energetic photons that destroy their genetic material. It's the reason that the sun causes clothes to fade. Ultraviolet light affects human tissue, too: it can cause cataracts and skin cancer.

What Brenner realised was that a shorter-wavelength variety, known as far-UVC, might be able to damage microbes without harming humans. It would penetrate the thin layer of dead skin cells that clings to the human epidermis, but it would go no

* Methicillin-resistant Staphylococcus aureus.

further. In theory, one could illuminate a hospital room with far-UVC light, killing every bacterium the light touches but doing no harm to humans. The innovation could therefore be the modern-day equivalent of the discovery of antibiotics – a super-weapon in the war on illness. MRSA and the other superbugs would have no defence against this invention.

From 2011, Brenner and his colleagues began further study of far-UVC light. Their goal was to lay the scientific groundwork for widespread adoption of far-UVC lighting in public spaces such as schools, nursing homes, trains and public transport.

Then arrived a menace that made Brenner's work even more urgent. As he realised in the spring of 2020, his invention has significance beyond antibiotic resistance. The COVID-19 pandemic had arrived and the death toll hit six figures, seven and then eight – numbers that rivalled the 10 million who die each year from superbugs.[2] COVID confined billions to their homes and wiped trillions from the global economy. Brenner and his colleagues, focusing on their potential superweapon, demonstrated that shining three millijoules of energy per square centimetre of surface – a paltry amount of energy – deactivates about 90 per cent of viruses after eight minutes and 99.9 per cent of viruses after 25 minutes.[3] As Brenner said, understatedly, of far-UVC light: 'It's very efficient at killing coronaviruses.'

Stimulated by Brenner and the pandemic, other scientists began to work on far-UVC light. The science remains a work in progress, for we still do not fully understand the far-UVC's efficacy, safety and environmental toll. At wavelengths below 240 nanometers, it produces ozone gas (O_3), which can damage the lungs when inhaled. Ozone gas can be dealt with by air purification, but it is expensive to install and maintain new lighting systems; far-UVC lamps cost hundreds of pounds.[4] An additional limitation is that a far-UVC lamp has to be in 'sight of' a surface in order to kill microbes on it. The challenge now is to make the lamps cheaper, and to study their effects further.

We will need tools like Brenner's far-UVC if we are to withstand a frightening new age of pandemics. Throughout history, pandemics usually came from the animal world. That's where we got bird flu, swine flu and HIV, among others. In each case, a pathogen that was virulent within a species of animal happened to mutate in a way that allowed it to cross over to humans. This is what biologists call 'zoonotic transmission' – when plagues successfully hurdle the low wall that separates the animal kingdom from our own. The squalid conditions of our factory farms, in which unhealthy animals are massed together, make such a hurdling all the more likely.

Zoonotic transmission, though, seems quaint when one considers how else a modern pandemic might arise. In the modern era, we have acquired abilities that our antecedents would regard as God-like. We have not only identified the microscopic pathogens responsible for diseases, but we can also manipulate them in a laboratory, even edit their DNA. As with the splitting of the atom, and as with the creation of artificial intelligence, the advancing of the scientific frontier has brought both treasure and danger. In 1980, we eliminated smallpox, hitherto humanity's most bloodthirsty enemy* – a magnificent achievement for which a great deal of the credit goes to Viktor Zhdanov, deputy health minister of the Soviet Union. Yet even as Zhdanov was campaigning for quarantining, isolation and vaccination, his Soviet party colleagues were designing factories that would pump out anthrax† and smallpox by the ton.[6] In huge vats was stored the infectious material that, if launched at the United States, could have killed millions.

In the end, the Soviets' anthrax-manufacturing facilities were dismantled, but some countries still defy the UN treaty that, since 1975, has outlawed such programmes. American intelligence suggests that Russia, North Korea and China are members

* It caused hundreds of millions of deaths in the 20th century alone.[5]

† See this book's introduction.

of the small but shameful club of countries that have active biological weapons programmes. These countries probably all harbour even the smallpox virus – as does the United States.

Our best defence against these weapons is not so much modern medicine, important though it is, but the fact that disease does not discriminate by nationality. Inflict a plague on another country and there's a good chance it'll come back to you. This makes governments reluctant to deploy disease as a weapon; people in glass houses shouldn't throw boomerangs. There are certain groups, however, that have fewer such qualms. In 1980s Oregon, for instance, the Rajneesh cult deliberately spread salmonella and considered spreading the AIDS virus.[7] In 2001, a senior biodefence researcher terrified the US by putting anthrax in the mail, killing five people.[8]

Of course, deliberately caused outbreaks are not a new idea; in 1346, for instance, the Mongols catapulted plague-ridden corpses over enemy ramparts at the Siege of Caffa.* But modern ingenuity makes deliberately caused pandemics a greater threat than they used to be. We have a better understanding of how to isolate deadly pathogens; we can better distribute them; and, as of more recently, we have acquired the ability to make them even deadlier, for scientific advances have made it easier to edit the genes of organisms large and small. DNA synthesis is the creation of new DNA to slot into an existing genome, and its cost has fallen dramatically. This is a trend that has democratised the ability to create potentially devastating pathogens. In 2001, a team of Australian scientists were attempting to create a viral contraceptive for mice when they accidentally produced a version that killed all the animals in nine days.[10] And, in 2017, a Canadian scientist recreated horsepox, an extinct virus with a very similar genome to that of smallpox.[11] Imagine the damage a rogue state could do if it made similarly inventive use of existing genetic material.

* So said a contemporaneous account, anyway.[9]

Since the horsepox experiment of 2017, technology has democratised further. Synthetic DNA can be ordered online, with orders not always screened for dangerous combinations. DNA synthesisers themselves can be bought online for €25,000, further lowering the obstacles to the creation of a pandemic.[12] These benchtop synthesisers, as they are known, are currently rudimentary and not yet capable of printing the DNA of a virus. But their successors, a global security think tank reported, 'could enable a user to create pathogen or toxic DNA without detection'.[13] The astronomer royal, Martin Rees, has written of his worry of a 'catastrophic natural disaster' being easily concocted by 'individuals with the mindset of the people who now design computer viruses'.

That makes two avenues by which pandemics can arrive: from nature and out of malice. There is one more route. And, again, what arrives along that route could result in a much higher death toll, in a given year, than do superbugs. Perhaps it has already happened.

Earlier I referred to the Russian anthrax programme. A country might never deploy its cache of biological weapons, but the simple fact of their existence exposes the world to risk. Indeed, dozens of Russians died in 1979 when anthrax spores leaked from a military facility in the Urals.[14] The leak was accidental, but accidents are a fact of life. It is a fact we must bear in mind when we consider the countless pathogens that are, at this very moment, kept alive in scientific laboratories in dozens of locations around the world. Virologists and other scientists usually have innocent intentions; their research is intended to help us understand pathogens and, ultimately, to better defend ourselves against them.[15] A negligent laboratory, though, could accidentally expose the world to pathogens more dangerous than those found in nature. The laboratories that conduct the most hazardous work are in the category of 'biosecurity level 4' (BSL-4) and there are at least 69 such labs in operation, under construction or in the planning stages.[16] The work undertaken in these laboratories sometimes includes 'gain-of-function' research – experiments whereby scientists take an existing pathogen and make it deadlier or more virulent. (The term 'gain-of-function' arose in 2012[17] to describe work in which avian flu was altered such that it could infect ferrets via air droplets.[18])

Having learned of the danger of gain-of-function research, you might seek to reassure yourself by imagining the safety record of these modern laboratories to be impeccable. Surely, you tell yourself, modern Western facilities must be better than their rusting Soviet cousins.

Alas. Accidental releases happen far more often than they should. Since 2000, in the US alone, live anthrax has been shipped to dozens of labs;[19] a researcher gave himself bird flu by accidentally pricking his own finger with a needle loaded with the virus;[20] six vials of potentially deadly smallpox were left lying around in a government agency office and remained unaccounted for at some point from the 1960s onwards and weren't found until 2014;[21] and, in 2009, a professor of biology was killed by a disease we thought we had seen the back of: the Plague.[22] And in the UK, in 2001, nearly 6.5 million sheep, pigs and cattle were slaughtered in order to control an outbreak of foot-and-mouth disease.[23] Never again, said politicians and farmers, as the carcasses of diseased and healthy animals alike were loaded onto pyres and incinerated. Yet, in 2007, another outbreak was caused when the virus slipped out of ageing pipework at a research fatality. Two laboratories used the facility; as they haggled over who should pay for new plumbing, heavy rains overwhelmed the drainage system and washed the virus into the open.[24]

Was COVID-19 what virologists call a lab leak? At the time of writing, American spy agencies remained divided.[25] Early in the pandemic, the virus was said to have emerged at the Wuhan Seafood Wholesale Market. It seems noteworthy, though, that the Wuhan Institute of Virology, which is on the other side of the Yangtze River, was doing dangerous research into coronaviruses.

Perhaps it was a momentary lapse in laboratory safety protocol that cost the world $12.5 trillion and counting.[26] But even if we put COVID-19 to one side, it seems clear that, given the existing rate of lab leaks, we can expect more leaks in future.

In short: though humanity has got better over the centuries at fighting pathogens, the chance of a self-inflicted pandemic – whether through a lab leak or a deliberate attack – has sharply risen. As a UK government review warned in 2021, the arrival of a novel pandemic by 2030 is a 'real possibility'.

Let's imagine that this 'real possibility' turns out to be a reality. The World Health Organisation (WHO) has a name for such a new, infectious condition: disease X. And in this case, as always, disease X arrives by stealth. Sometime in October 2030, several residents of the Siberian city of Novosibirsk complain of illness. They are feverish and fatigued, and when they present themselves to doctors, they are told they have the flu. The patients return to their families, but some of them deteriorate. They begin to suffer from memory lapses. A few become aggressive.

Within the following fortnight, more and more patients present themselves to doctors. The local hospital diagnoses a case of encephalitis, in which the brain becomes inflamed. Then another. And another. The local doctors, increasingly concerned, no longer view this outbreak as a seasonal flu. They report it to health authorities, who request a blood sample so that they can sequence the virus' genetic code. The sample is sent to the Centre for Algorithmic Biotechnology at St Petersburg State University, where researchers discover it to be a virus that combines the genetic material of both an arenavirus and a lyssavirus. Arenaviruses, such as lassa fever, are adept at suppressing immune responses; lyssaviruses, such as rabies, can cross the blood-brain barrier to cause nightmarish neurological symptoms. This new virus is a mix of both.

Back in Novosibirsk, the city is already in a self-imposed lockdown. This is the third-biggest city in the country, and hundreds are dying. Their neighbours, recalling that COVID was transmitted predominantly via aerosols, dig out their masks, but people still get sick. Around the world, footage is circulated of sufferers trembling, speaking in slurred fashion, shouting at each other and collapsing. In hospital, they assault healthcare workers. When death finally comes for those infected, it is through their immune systems overreacting to the viral load. As a result, they suffer septic shock and multiple organ failures. At victims' bedsides, family members behold mottled skin, swollen faces and shallow breathing. It is cardiac arrest that usually administers the killing blow. Bedside visits are soon curtailed.

The scientists at St Petersburg realise immediately that this disease is far worse than COVID-19. Under pressure from China, the Russians allow WHO officials into Novosibirsk.

I stress that no such virus is known to have ever been made, by Russia or anyone else. Arenaviruses and lyssaviruses come from different viral families and would not be easily mixed. To would-be terrorists searching for a failsafe recipe for disease X: this is not it. (For obvious reasons, pandemic specialists prefer to keep such recipes from public view.) The characteristics and origin of my hypothetical virus, then, are less important than its severity; this thought experiment rehearses a worst-case scenario. How, when faced with a virus worse than anything we have found in nature, will our species fare?

It depends on the lessons we learned from the COVID-19 pandemic. Toby Ord is the Future of Humanity Institute veteran who catalogued catastrophic risks in *The Precipice*. That book was published in March 2020, just as COVID was erupting. Reflecting in 2024 on the extent to which the world had become any safer, Ord wrote that post-pandemic spending on biodefence, at least in the UK and US, had been 'disappointingly weak'.

I am going to charitably assume that this weak level of spending does not imply that the world, come 2030, will be unready for the emergence of disease X. I specified the year 2030 partly because that was the year referred to by the UK government review into pandemic risk and partly because it gives us enough time to develop better methods of staving off dangerous pathogens. By 2030, in this hopeful scenario, David Brenner and company have long since commercialised far-UVC lighting. Brenner lamps are installed in airports, which means that when travellers expel aerosols when talking, coughing and sneezing, the airborne virus particles suffer grievous losses. Brenner, watching the news from his penthouse, is relieved to see that the spread is slower than it could have been.

A weakness of far-UVC lighting is that, in order to be effective, it must be, as it were, in sight of pathogens. Unless we install recessed floor lights in every room of every building, and keep them lit 24/7, far-UVC lamps will remain an exciting addition to our pathogen-fighting arsenal but not a cure-all. Our troubling new arenavirus-lyssavirus spreads, like many other viruses, not only via air droplets but via surfaces.

There is a name for the objects or inanimate surfaces that carry infection, and it is 'fomite'.* Most square centimetres of our built environments transmit next to no disease; think of walls, which we rarely so much as brush in passing. But some are touched with extraordinary frequency. Fomites can include elevator buttons, door handles, light switches and clothing. All these objects harbour pathogens – not just viruses, but also bacteria and fungi – which our tactile species then industriously and unwittingly spreads around. Some surfaces, because they are porous, are especially hospitable to pathogens. If you recently bought this book from a sneezing cashier, or if it's been passed to you by a family member unknowingly harbouring an illness, then you, too, might be handling a fomite.

In 2023, on behalf of an American biosecurity think tank, I searched for neglected ways in which to make the built environment – our homes, shared buildings, towns and cities – less

* The name comes from the Latin *fomes*, meaning 'touchwood' or 'tinder'.

prone to fomite transmission. We already have a widely used method of sanitising surfaces, of course, in a variety of bleach-based products. Bleach belongs to a family of disinfectants that often contain sodium hypochlorite, a chemical that, when applied to any of a wide range of substances, causes oxidation: the loss of electrons. At a cellular level, oxidation is brutal. It renders proteins non-functional, inactivates key enzymes, damages DNA and causes cells and viruses to break apart. Bleach is used worldwide in hospitals, on public transport and in homes. It is cheap and easy to use, but it is also a blunt tool with its own drawbacks.

The overuse of bleach, as with most biocides, contributes to antibiotic resistance.[27] This is because oxidation, brutal though it is, can be partially resisted. When the concentration of bleach is low, or when bacteria are regularly exposed to sodium hypochlorite, the cells improve the ability of efflux pumps, which I mentioned earlier in the context of antibiotic resistance, to expel toxic substances. These toxic substances, from the perspective of a bacterium, include antibiotics.

Additionally, bleach and its fumes are toxic – you, too, are made of substances that can be brutalised by oxidation – and are associated with conditions such as asthma.[28] It also needs to be reapplied at frequent intervals, and damages absorbent surfaces. Using bleach to kill the odd cold virus is like using acid rain to put out a fire.

Can we do better? The Egyptians wouldn't have understood the chemistry,* but they knew that copper was a wise material

* Copper is a transition metal, the kind that has the propensity to lose electrons. This happens to copper when it is exposed to air and moisture. When copper atoms lose electrons, they become ions and the positive charge of the copper ions allows them to chemically interact with atoms they come into contact with. As with the comparable phenomenon of oxidation, this causes huge physical disruption to microbes: a wind of chaos that denatures proteins, degrades genetic material and breaks viral envelopes and cell membranes.

from which to make a drinking vessel.[29] In hieroglyphs, the ankh symbol refers to both copper and eternal life. The metal was used for its antimicrobial properties by the Phoenicians, ancient Chinese, Greeks, Romans, Aztecs and others. In the 18th century, copper was used to prevent the build-up of algae and other biological matter on ships' hulls. Copper, and its alloys, are still used as antimicrobial agents in the modern world, and renewed medical interest in copper has led to reams of supportive research.[30,31] Although some microbes can develop a certain level of tolerance to copper, just as they can with bleach, copper is much less toxic to humans and gives off no fumes.

Should we therefore replace every high-touch, non-porous surface with copper? On behalf of the biosecurity think tank, I did a back-of-the-envelope calculation of the cost of replacing every light switch in American public-sector buildings: schools, hospitals, prisons, government buildings, and so on. Given that there are around 140,000 in the US,[32] we can safely assume that the total number of American public sector buildings is greater than 500,000. Multiply that by the 100-odd light switches that you'd expect per building, at a cost of roughly $20 (£16) each, judging by existing prices, and assume that the labour costs are a conservative $25 (£20) per switch. That's a billion dollars (£780m) on parts, plus $1.25 billion (£980m) on labour – and that's only for light switches.*

In the future, we might copy nature's more mechanical methods of slaying microbes. The wings of cicadas, for instance, are antimicrobial by dint of the infinitesimally small spikes with

* Silver, which is also anti-microbial, is no cheaper. Those truly interested in the antimicrobial merits of this metal can refer to some additional research that I conducted as a younger man, when journalistic misadventure forced me into wearing the same 'self-cleaning' pair of underpants for five days. If you'd prefer to avoid the gnarly details, know that silver's antimicrobial effect is enhanced by humid environments, but that, in this case, it stopped some way short of perfection.

which they are covered. These tiny structures – nano-pillars – have been replicated by scientists, but they are prohibitively expensive.

At this stage of my research, it all seemed futile. Self-sterilising surfaces are a promising technology, but there is no world in which they will be immediately and universally rolled out. Copper-based sprays are promising, too, but still relatively expensive.* Just as intractably, a coppery finish might not be something that consumers want on every single surface. What we accept in hospital fixtures, for instance, we might not want on our phone cases.

It was then that I learned that, in 2023, researchers at the University of Birmingham announced the invention of a sprayable coating derived from cellulose.[33] This hydrophilic coating kills pathogens by drying out the respiratory droplets in which they lurk. Viruses and other microbes then become ensnared in the cellulose matrix and don't spread as readily when touched. A material coated in the cellulose spray dries about 50 per cent faster than the same volume of liquid on a glass surface. In the study, glass was coated with cultures containing infective particles of SARS-CoV-2 particles. When the cellulose spray was applied, the size of the culture was reduced almost three-fold after five minutes. After ten minutes, it was almost entirely removed.

The research was led by Zhenyu Jason Zhang, professor of soft matter engineering. Zhang told me that he and his team had taken their cellulose from eucalyptus pulp, but that they would also be interested in trying fibres extracted from pineapple. 'When you eat pineapple, you very likely feel the zinginess. That's because of the enzymes within it. They disrupt the protein films of cells in your oral cavity.' For all the talk in scientific

* An American start-up that makes such a spray has gained a licence to sell its product to customers in the consumer, commercial and healthcare sectors. Coatings will cost between $50 and $60 per square foot.[34]

circles about our having plucked the low-hanging fruit, pineapple might be the exception.

If this cellulose spray ever makes it to market, it would make for a product that's cheap, non-toxic and transparent, minimising its aesthetic intrusion. Together, these qualities offer a trifecta that neither bleach nor copper can match. The spray comes off when wetted,[35] as disinfectants do, which means that we shouldn't use it for drinking beer with ancient Egyptians. Nevertheless, pending more research into its relative efficacy and microbes' ability to adapt to it, cellulose spray is a technology to keep an eye on.

And perhaps, in our 2030 disease X scenario, cellulose has become a fairly widespread alternative to bleach. The arenavirus-lyssavirus has a long incubation period, which means that mask-wearing takes a while to catch up, but humanity's passive defences have been silently doing their part. In airports, schools and hospitals, the virus particles are being killed in the air (zapped by Brenner lamps) and on land (desiccated by Zhang sprays).

At sea – or, rather, in our plumbing – the particles are facing similarly unfavourable odds. For many years, humanity has chemically treated its wastewater in order to kill pathogens. But a newer innovation is the scanning of wastewater for pathogens. In the US, the COVID pandemic prompted the creation of a scheme whereby sewage treatment plants helped monitor inbound wastewater for COVID, providing valuable data about local outbreaks. But, today, scientists are in the early stages of developing what's called metagenomic sequencing: an analysis of a mass of genetic material, sequencing every genome present. Take a sample of wastewater, air or anything else that might contain pathogens, and your analysis will throw up scraps of RNA and DNA. With the right software, and enough computational power, you can work out what those scraps came from and what characteristics that microbe might have. If your metagenomic sequencing is sufficiently sophisticated, you might be able to spot a new pathogen before there is a visible outbreak.

For now, the process is difficult and expensive. It will be some time before everything that comes out of the Armitage Shanks factory is equipped with inbuilt metagenomic sequencing apparatus. But the cost of sequencing a human genome has fallen a million-fold since the turn of the century[36] – a development that signals great potential in the related technology of metagenomic sequencing. In November 2024, the British government announced plans to set up the world's first real-time genomic surveillance system, spanning at least 10 NHS sites.[37] (Genomic surveillance, as opposed to metagenomic surveillance, targets specific pathogens of interest rather than sequencing every genome present.) The same should be done elsewhere. According to one proposal, a system of genomic surveillance could be set up in 200 busy hospitals in the US for much less than $1 billion, and it could be done within a year. This would provide a priceless source of data on new outbreaks.

Not enough is being done to accelerate this technology and install it in as many places as possible. But we can hope that, in 2030, global health authorities notified of the arenavirus-lyssavirus will be able to set up targeted surveillance of wastewater. (Such technology has already been deployed in order to monitor the spread of such pathogens as poliovirus[38] and, of course, coronavirus.[39]) Knowing of the new outbreak, the cities served by Novosibirsk airport will be particularly watchful, as will those with high levels of air traffic. The more precise the monitoring, the more effective the local measures to slow the spread of the virus.

British veterans of the COVID-19 pandemic will recall the inordinate faff that was the test-and-trace system, whereby members of the public had to take frequent self-administered tests and report close contact that might have involved an infection risk. Technologies such as far-UVC lamps, self-sterilising surfaces and metagenomic sequencing have some major advantages over test-and-trace. For one, they do not ask as much of the public. They are pragmatic innovations that take account of human nature. Additionally, as Ord points out, they address everyday infectious diseases such as the flu.*

At some point, though, we will encounter a pathogen bad enough to force the public to change its behaviour. The arenavirus-lyssavirus, the subject of our imagined 2030 scenario, is one such pathogen. Across the world, the public has begun to shutter itself. There is predictable chatter about the virus being a hoax, but most people realise that this virus is worse than COVID-19. Its long incubation period has ensured its stealthy spread: first to the cities with the busiest air traffic hubs – Singapore, Dubai, Hong Kong, London and so on – and then into other areas of high population. Two months have elapsed since the virus first infected a human. Cases are surging. By now, scientists have made several analyses of the virus. It is highly unnatural in the way it combines genes from two distinct types of virus. Outside Russia and its allies, few experts contradict the view that this pathogen was made in a laboratory.

For now, though, the virus' origins are less important than its treatment. Governments are scrambling to add capacity to their hospitals. COVID-era ventilators are brought out of storage. Vaccines manufacturers are harangued night and day. It took six weeks to create the first COVID-19 vaccine, but a year to complete testing and get shots in arms.†[40]

This magnificent scientific effort minted several new members of the anti-catastrophe league, including Katalin Karikó. The Hungarian-born biochemist had laboured for years on a vaccine technology that used messenger RNA,‡ or mRNA for short. As

* Ord is writing about far-UVC lighting and metagenomic sequencing. He does not include self-sterilising surfaces, a less consequential technology, in his discussion. But the analysis applies equally.

† Compare this to the 10–15 years of research that is usually required to create a vaccine.

‡ RNA is ribonucleic acid. Think of it as DNA, but for viruses.

Karikó knew, mRNA tells cells which proteins to make. As Karikó theorised, mRNA vaccines could tell cells to make antibodies specialised to fight specific pathogens – if only she could work out how to stop the body's immune system from destroying this invasive mRNA. Grantmakers saw little potential in Karikó's idea. At the University of Pennsylvania, Karikó was refused promotion. By her account, she was called 'difficult' and was denied basic laboratory equipment. 'Usually, at that point, people just say goodbye and leave because it's so horrible,' she later recounted.

After a cancer scare, Karikó considered leaving the university.[41] She was not, apparently, 'of faculty quality'.[42] But she and a colleague, Drew Weissman, found a way of adapting mRNA such that it could slip past the body's immune system guardsmen. Their research, published in 2005, received little immediate recognition, but became crucial to a new generation of vaccines that used mRNA technology. By the time COVID hit, Karikó had left academia to help Moderna, a biotechnology company, create these vaccines. Moderna's vaccine was highly effective and thus helped to save the lives of millions.[43] It is safe to say Karikó was, in fact, of faculty quality. She would later receive not only a Nobel prize, but also an adjunct professorship at Penn.

It would be facile to conclude that the moral of Karikó's story is that one ought never to give up. Most projects can be abandoned with little consequence. But if it is true that Karikó's treatment was a result of her failure to secure grants, then we should question the criteria by which grants were awarded. We should also question the university staff's apparent outsourcing of its judgement to those grant-makers. There is little scientific research more important than the kind that helps humanity avert catastrophe. If a particular line of research seems unlikely to pay off, that unlikeliness should be weighed against its high potential.

We return to 2030. Perhaps today's entrepreneurs will have succeeded in creating better defences against viruses: products such as nose sprays that inhibit viral particles' reproduction and transmission.[44] We can also hope for improvements in the quality and distribution of antivirals – drugs that are not vaccines but can be effective against viruses. They tend not to be as precisely targeted as vaccines, but can buy us time.

And, if we are lucky, vaccine development and distribution will be quicker than it was in the last pandemic. The Coalition for Epidemic Preparedness Innovation, a non-profit that supports work on vaccines, is currently leading what it calls a '100 Days Mission'. Its aim is to enable widespread vaccination within 100 days of the identification of a viral threat. A promising development, in this regard, is the rolling-out of small and widely distributed mRNA manufacturing plants.[45] These modular units, some the size of shipping containers, will make it easier for the world to rush out vaccines by the million. There should be no repeat of Boris Johnson, then the prime minister, wondering whether he would have to send in special forces operatives in order to retrieve UK-bound COVID-19 vaccines held up in the Netherlands.*

In our lyssavirus-arenavirus scenario, millions could be infected by January 2031, three months after the first patient lumbered into hospital. The world's streets are empty. A quiet terror reigns. But in this quiet can be heard the whinnying of the cavalry, for the vaccines are on their way.

But this particular virus is so horrible that it is difficult for society to maintain its most basic functions. Just as in the COVID pandemic, certain lines of work demand in-person attendance: agriculture, energy, healthcare, policing and so on. But the death rate of the lyssavirus-arenavirus is palpably higher. In 1986, Soviet officials could call on those dutiful helicopter crews to absorb mammoth doses of gamma radiation while

* Perhaps a fanciful claim on the part of Johnson, who had a book to sell. It's true, though, that there were significant logistical problems with the roll-out of COVID vaccines.

dropping clay on the Chernobyl reactor fire. In a free society, though, that is too much to ask of those workers without whom society cannot function. They need equipment that will protect them from even the most infectious outbreak. The COVID pandemic stimulated improvements to personal protective equipment (PPE), but the gear remains expensive. 'A clear market failure is happening,' said Tom Milton, CEO of an engineering consultancy that has conducted thorough research into the possibilities of next-generation PPE.

Governments can stimulate this market by commissioning gear that can be either stockpiled or used in existing healthcare situations. This equipment must ensure that workers can do their jobs without fear of acquiring an infection. PPE must be airtight, cheap, comfortable and even, some have argued, fashionable. 'You can see people's lips,' said Milton of powered air-purifying respirators, which fit over a user's head and use a transparent screen to shield the face. 'The sound isn't muffled as much. So it's not just that it's much more protective, but also that it's much more useful. I would actually prefer this in a non-existential pandemic as well as an existential pandemic.'

Within a year of the start of the 2030 outbreak, the lyssavirus-arenavirus has loosened its grip on humanity. Its spread was slowed by far-UVC lamps and self-sterilising surfaces. Essential workers, protected by next-generation PPE, kept society's basic systems functioning. Then, mRNA vaccines arrived. Like COVID-19 before it, the lyssavirus-arenavirus has mutated into a milder form against which we have immunised protection. In the forest of disease, it has become a hardy perennial rather than an apex predator.

But it has killed millions of people. It has wiped trillions of pounds from the global economy. Despite the Kremlin's protestations of innocence, intelligence agencies view it as highly likely that the virus was a lab leak. Just as Wuhan is home to a virology research centre, so Novosibirsk is the neighbour of a similar institution: Vector. At the edge of the snowy wilderness sits this

Soviet-era research centre, formerly the home of a bioweapons programme. It is here that the lyssavirus-arenavirus is believed to have been created and from here that it managed to escape.

CHAPTER 10

THE WEST SIBERIAN HAT FACTORY – AND WORSE

'Chance favours the prepared mind'

Louis Pasteur

The arenavirus-lyssavirus does not exist, but Vector does. It is a vast research facility that was founded in 1974, the year before the Biological Weapons Convention came into force. The Biological Weapons Convention is an international treaty that prohibited biological weapons and the USSR was one of the signatories.

On the face of it, Russia complied with the rules of the treaty. Vector* had a putatively innocent mission: to develop biological pesticides for agriculture.[1] Its assistant scientific director was Lev Sandakhchiev, a man described by the Cold War chronicler David E. Hoffman as 'a compact, intense, chain-smoking scientist of Armenian descent'.[2] Sandakhchiev, Hoffman writes, offered young scientists salaries that were 50 per cent greater than they could get elsewhere. Come work for Vector! Make discoveries! Improve agriculture! The reality, as new recruits would later find out, was that Vector's huge campus concealed a rather less beneficent project. 'At the core,' writes Hoffman, 'it was a laboratory to discover new methods of using viruses to kill people on a massive scale.'

* Full name: the State Research Center of Virology and Biotechnology.

Sandakhchiev wanted his bright young scientists to learn how to isolate and manipulate DNA, create dangerous viruses and test them on animals. New viruses would then become part of a roster of biological weapons. Sandakhchiev wanted a new virus every month, and if a researcher didn't want to contribute, then the KGB would follow them for the rest of their life. And there would be no hope of a remotely attractive new job. It was Vector, the researchers used to quip darkly, or the 'West Siberian Hat Factory'. Few of them could muster the requisite appetite for frostbitten millinery.

The researchers saw their programme as a necessary evil – an attempt to match whatever the Americans were cooking up.[3] The US had indeed had a biological weapons programme that encompassed the mass production, in the Second World War, of anthrax and other pathogens. In open-air tests conducted in Utah, these pathogens were dispersed via bombs, bomblets and sprays.[4] The US developed bioweapons that could kill crops as well as humans, and persisted with the programme into the Cold War. But from 1969, at the instigation of President Nixon, the US unilaterally dismantled the programme. The Soviets, if they did but know it, were racing a phantom.

But they ran as fast as they could. As I wrote in the previous chapter, the world eliminated smallpox from circulation in 1980. But the US kept samples of it and the Vector scientists continued to experiment on it. Smallpox, though, turned out to be far from their most disturbing weapon. Encouraged by Sandakhchiev, a scientist called Sergei Popov discovered, along with his team, a method of inserting into microbes a gene that triggers the creation of myelin. Myelin is the fatty tissue that sheaths neurons, and the apparition of foreign myelin causes havoc. Failing to distinguish between friend and enemy, the victim's immune system would attack its own myelin. This would cause brain damage, paralysis and near-certain death.[5]

A key question for Popov was which microbe to attach his invention to. At Vector, and then at another laboratory at Obolensk, near Moscow, he and his colleagues began experimenting. One early substrate was smallpox. Later, Popov and his team tried a range of bacteria, including the variety *Yersinia pestis*, which causes bubonic plague.

An infection would have gone something like this. First, the substrate would cause an initial infection from which a victim might experience a temporary recovery. Then the auto-immune response would begin. 'As a weapon, the thing would be absolutely untreatable,' said Popov. 'Absolutely untreatable, because first of all it may come as a surprise after the initial disease has gone away, the person may recover completely. And then the new wave of disease would be the death response ...'[6]

Popov worked on these appalling chimaeras through the 1980s and beyond. Only in 1992, when the USSR had fallen, did he leave Obolensk. He defected to the West and lives, to this day, in the US.

The staff of Vector and Obolensk had left the Soviet bioweapons programme all but unguarded. Andy Weber, the American diplomat whom I introduced at the beginning of the book, had been tasked with tracking down the Soviets' nuclear weapons, but by 1997 he was on the hunt for their bioweapons. At a scientific conference in Kirov, five hundred miles east of Moscow, Weber charmed a group of Obolensk scientists into inviting him on a trip to a banya, a traditional Russian sauna. As the men chatted in the heat, Weber learned that the Iranians were trying to find out from the Russian scientists how best to make biological weapons. Hoffman writes that a Russian scientist later showed Weber a business card that had been given to him by the Iranians. Weber immediately recognised the name on the card: an alias used by the Iranian intelligence services.

Concerned, Weber approached one of the few men who might know where most of the bodies were buried.* It was Lev

* I mean this semi-literally; Vector even had its own graveyard. Here, in a zinc-lined tomb, lay the body of a scientist who had accidentally injected himself with Marburg virus.

Sandakhchiev whom Weber contacted – the scientist who had recruited Popov, who had gone on to direct Vector.

Following the fall of the USSR, Sandakhchiev and his colleagues had gone unpaid for months at a time. One way for Sandakhchiev to rescue the laboratory and his scientists might have been through cooperating with the Iranians. But the other, it seemed, was to admit to the West the existence of the bioweapons programme. In 1993, at an international meeting of scientists, Sandakhchiev had contradicted his compatriots' denial of this programme's existence, startling his colleagues by shouting the Russian equivalent, as his obituary tells it, of 'Let's cut the crap!'

Weber asked Sandakhchiev to show him Vector. Here, as elsewhere in the USSR, Weber beheld a level of disorder that, given what the site was harbouring, would have been laughable had it not been so dangerous. 'It was clear the place was just a wreck,' Weber told Hoffman. 'Crap all over the floors, the equipment was in terrible shape.'

The Russians were slow to trust the Americans, but Weber managed to convince Sandakhchiev that he could help save Vector. With the help of his colleagues in the State Department, Weber brokered an investment worth \$3m. It was enough to persuade Sandakhchiev to cut ties with the Iranians and to help Weber safeguard what was left of the bioweapons.*

After making this deal, incidentally, Weber learned what some of the Soviets' agriculture-related research had secretly been directed towards: wiping out the American wheat supply, while ensuring that Russian crops would grow in a nuclear winter.[7]

* Sandakhchiev presents a compelling case for both inclusion in the anti-catastrophe league and being blackballed from it. For what it's worth, the Pentagon sent a wreath to his funeral.[12]

Vector remained a functioning research institute, even retaining a stock of smallpox. As a BSL-4 laboratory, it is registered with the Biological Weapons Convention and has been inspected several times by the World Health Organisation, generally providing little cause for worry.[8] In 2019, an explosion caused a fire at Vector, prompting worries that live smallpox might escape, but it appears that nobody was infected.[9] Since invading Ukraine, Russia has not withdrawn from the Biological Weapons Convention, though the US state department accuses Russia of maintaining an offensive bioweapons programme – that is, one designed to develop or maintain bioweaponry, rather than simply to defend against it.[10] Since the war began, there have been no international inspections of Vector.

Let me be clear: my arenavirus-lyssavirus scenario is inspired by Soviet pathogen-splicing rather than by any evidence that the current Russian regime has produced such a chimaera. Any such lab leak would be a worst-case scenario. A widely cited analysis puts the likelihood of a COVID-scale pandemic at about two per cent in any year;[11] an arenavirus-lyssavirus outbreak could be more damaging, but it is much less likely. But these odds do not entitle us remotely to relaxation. For all the innovations I referred to in the previous chapter, for all that new technology that can help us mitigate incoming pandemics, it would be much better if we could prevent them from occurring at all. And to prevent them from occurring, humanity must pay close attention to those BSL-4 laboratories I mentioned in the previous chapter – Vector and its dozens of counterparts across the globe.

Enter one of these BSL-4 laboratories and, if their safety practices are up to scratch, you will first step through an airlock.[13] Then you will have to wear a voluminous suit that is continuously supplied with filtered air. If, in the course of your work, you pierce the suit, air will rush out rather than in; this is a result of its artificially high pressure. On your way out, you will have to take a chemical shower designed to kill any pathogens.

Many more rules govern these facilities. In general, they are managed with care.

But there are so many of them. And not every laboratory is being built in countries renowned for optimal biosafety. All in all, it is a troubling situation. In 2023, I spoke to Filippa Lentzos, an academic and biosafety expert who leads a project that keeps track of all these new labs, comparing their safety levels and recommending good practice. Only a few weeks before we spoke, members of an armed faction fighting in the Sudanese civil war had seized a laboratory that held measles an cholera pathogens.[14] There had also been a flow of Russian disinformation about biolabs in Ukraine, as well as the continued discussion of the origin of COVID-19. 'It's a lot,' said Lentzos.

The increase in the number of BSL-4 laboratories, she said, involved countries without experience of safely maintaining these facilities. Some of these countries do not score highly on measures of biosafety; for instance, Gabon and Côte d'Ivoire scored a single point out of a possible 18. 'There are now countries that have built these high-containment labs without having these oversight systems in place, and without having that experience of working safely, securely and responsibly with these pathogens. So that is a concern.'

There is little hope of countries politely acquiescing to requests to shut down their laboratories. Lentzos and her colleagues, more pragmatically, have called for countries that host labs of BSL-3 or greater to abide by a common – and high – standard of biorisk management.[15] (Fortunately, one already exists, courtesy of the International Standardization Organization (ISO).[16]*) As they do in the sphere of nuclear arms control, countries could inspect each other's facilities. Lentzos

* What else, you might be wondering, has ISO standardised? Examples include the brewing of tea, the correct designation of the term 'organic' and the orthography of the Klingon language.

stressed to me that while a pandemic is always possible, she nevertheless sleeps soundly. 'There are risks of this happening, but they are low.'

So that's one way of reducing the likelihood of a pandemic: campaigning for biosafety. But there are other levers to press. Later in 2023, I spoke to Jaime Yassif, who is vice-president of a Washington DC-based non-profit that aims to reduce the risk of pandemics. The non-profit, NTI Bio, grew out of another non-profit, the Nuclear Threat Initiative (NTI). NTI's greatest hits include a key role in the conversion of 2,900 kilograms of highly enriched uranium reactor fuel, another legacy of the Soviet Union, into a less dangerous form of uranium.[17] NTI Bio aims to do similarly important work in biosecurity.

Yassif had begun her academic career with a biology degree, eventually taking work in the field of nuclear security. This was the early Noughties, and although the 'end of history'* had been ended in turn by 9/11, relations with Russia were much better than they had been in the 1980s. Sandakhchiev and Popov were no longer overseeing the creation of nightmarish new viruses, while the prospect of engineered pandemics was no longer giving diplomats much pause. But Yassif could see that synthetic biology was beginning to take off. 'It seemed pretty clear,' she told me, 'that biology was going to be a really important field for the 21st century.'

Her forecast was correct. As Yassif took increasingly senior biosecurity roles in government and non-profits, it became increasingly easy and cheap for scientists to assemble dangerous new pathogens in labs. At the same time, the number of labs handling the deadliest pathogens has multiplied.

How does one handle all these worries? 'I would actually say I'm pretty analytical about it,' says Yassif. 'I don't think I could work effectively if I was very emotional.' She refers to the human devastation caused by COVID and the bombing of

* To use Francis Fukuyama's infamous phrasing.

Hiroshima in 1945. 'You can't lose sight of that. But when I'm in my professional mode, I'm thinking about it as a system.'

And it is systemic solutions for which she and NTI Bio advocate. There is a ton of work that needs to be done to develop new and innovative and practical approaches to meaningfully reduce risks. Interventions that have teeth, that actually change the game.

One such intervention is the introduction of software to help DNA synthesis providers – the companies that sell scientists the genetic material from which they make new organisms – cheaply and easily screen customers and their orders. (You'd hope that alarm bells would ring if a Mr Kim, from the Democratic People's Republic of Korea, ordered smallpox DNA to Pyongyang.)

Another is the development of a diplomatic and scientific procedure by which experts, with UN backing, can investigate the source of new outbreaks.[18] The Joint Assessment Mechanism, as this diplomatic apparatus is known, would resolve ambiguity over whether an outbreak is emerging naturally – or is the result of a deliberate or accidental release. The prospect of such scrutiny, it's hoped, would prompt governments to behave responsibly. But it is a difficult time to be advancing policies even as sensible as this one; with the Trump administration pulling the US out of the World Health Organisation, the world's ability to collaborate on pandemics appears diminished.

Yassif is measured and professional. Where some lines of anti-catastrophe work reward spy-like adventuring (Andy Weber) or moments of dazzling technological innovation (Paul Brenner), hers demands patience, steeliness and realism. Weber told me that part of the pleasure of his job was the sense of being part of 'the biggest team in the world' – the US government. In the wider, looser team of people who are attempting to avert catastrophe, an extraordinary array of talents is represented. Those diplomatic talents are some of the most important.

And they will be severely tested over the coming years. The explosion of artificial intelligence will be an even more significant development than the explosion of synthetic biology. It will touch almost every human activity, including those that relate to dangerous pathogens. As Yassif and her colleagues have argued, AI will offer 'tremendous benefits' to bioscience and engineering.[19] It will support the rapid development of vaccines and therapeutics, they wrote, while enabling the development of new materials.

As an example, take proteins. These are the organic materials that execute the orders of DNA and RNA. They do so by folding into three-dimensional structures, which introduces complexity that makes them very hard to design. Google's DeepMind project, one of the leaders in the field of artificial intelligence, has made great progress in helping us model protein-folding. This progress invites the possibility of accelerated drug design, treatments for protein-related diseases such as Alzheimer's and the creation of – among many exciting organic products – stress-resistant crops, and enzymes that can digest plastics. 'Think of the cell as the factory floor,' Yassif told me. 'The protein is the machine on the factory floor that's doing the work. So, if we know how to design new machines that fundamentally change how living systems work, that's a pretty fundamental shift.'

Fundamental shifts are not always wholly good. Yassif is not particularly worried by large language models (aka LLMs), which make scientific knowledge more accessible to people with no formal training. More troubling is a new generation of bio design tools, a category that includes tools for engineering proteins. These tools, said Yassif, are 'consequential, because as they continue to advance, they may make it possible for sophisticated scientists to develop novel pathogens that are more dangerous than what we find in nature.

'We're not there yet, and there are significant technical hurdles, but we may get there soon.'

A final mode of threat is the ability of AI to automate bioscience research. It might currently take a human years to understand the scientific literature, formulate new hypotheses and test the hypotheses with work in a laboratory. If AI can do this on humans' behalf, it could make these processes happen many times more quickly. These AI-enhanced capabilities could be wonderful for the development of new therapies, but dangerous if turned towards nefarious purposes. 'That's a more distant issue on the horizon,' said Yassif, 'but one that we should watch. I think it's going to fundamentally change how scientists do biology.' In December 2024, scientists shared with the world a new peril.[20] They feared that mirror bacteria, which are a hypothetical, mirror-image version of existing bacteria, could cause catastrophe. Mirror bacteria behave in the same way that their original versions do, but they would sidestep the immune responses of animals, plants and humans. The scientists estimated that it will be a decade before we can create these mirror bacteria – but a dramatic acceleration of AI, it has been hypothesised in AI safety circles, might result in a shorter countdown.[21]

And these new bioscientific capabilities represent a fraction of the vast array of human endeavours to which superhuman intelligence will soon be applied. To avert catastrophe, we must avert pandemics, nuclear war, super-eruptions and all the other potential disasters I have chronicled thus far. But we will also have to find a way of harnessing the benefits of AI without exposing ourselves to terrifying new dangers. We are summoning a genie that we now must tame.

CHAPTER 11

THE RACE FOR ALIGNMENT

'Did I request thee, Maker, from my clay
To mould me Man?'

John Milton, *Paradise Lost*

'We are only undeceived
Of that which, deceiving, could no longer harm.'

T. S. Eliot, 'East Coker'

Why summon a genie at all? In fables, genies are powerful spirits who can make a character's dreams a reality. 'Your wish is my command,' as the saying goes. Aladdin's genie, for instance, frees our urchin protagonist from a cave, then conjures him a palace.

Sometimes, though, genies have wishes of their own. Beyond the story of Aladdin, these powerful beings can be devious and wilful. In 'The Fisherman and the Genie', a tale from the Arabian Nights, a fisherman finds in his net a sealed copper bottle. He opens it and out pours a vast, powerful genie.

This genie had been imprisoned for centuries by the King of the Genies and had vowed to kill whoever freed him. He informs the fisherman of his imminent fate, asking him how he would like to die. The fisherman thinks. Genies can be murderous, but they tend to be beings of their word. Asked by the fisherman to prove that he really had been imprisoned by the little copper bottle, the vast genie swoops back in. Our cunning fisherman stoppers the bottle and flings it back out to sea. The next fisherman might not be so lucky.

In the modern era, we do not summon genies. In artificial intelligence, though, we are creating something more powerful than ourselves, something that we wish to do our bidding. Like genies, AIs can be deceptive and dangerous; like genies, they follow instructions to the letter.

We might assume, naïvely, that those instructions – the wishes humanity puts to the genie – will always be benign. That still leaves two problems. The first is that the instructions we issue to the AI might be imperfectly aligned with the interests of humanity. The second is that the AI, being dedicated to the fulfilment of those instructions, might devise perverse sub-objectives. If you are truly dedicated to your goals, you will ensure that you gain as much power as possible and cannot be switched off.

When your wish is a super-intelligent AI's command, therefore, you had better be careful what you wish for. As Stuart Russell, author of the leading textbook on AI, puts it: 'This is essentially the old story of the genie in the lamp, or the sorcerer's apprentice, or King Midas: you get exactly what you ask for, not what you want.'[1] We don't yet have the problem of AIs that can turn everything into gold, but we are already seeing signs of deceptiveness and danger. In all, we face what researchers called the 'alignment problem': the challenge of aligning the behaviour of AIs with the goals of their operators.

Mid-century technologists foresaw this problem. As early as the 1950s, primitive proto-computers were the cause of anxiety about machine takeover. Even in the 1990s and the Noughties, though, these worries were largely theoretical. Eliezer Yudkowsky, Nick Bostrom and others discussed them with their fellow Extropians, but, at least at first, were more interested in the possibilities of wish fulfilment than genie-taming. AI could help us understand the universe, experience a far richer existence than our current biology allows, even defeat death.

Along the way, it will make its creators very wealthy. In the 2010s, investors began to put money into promising AI projects. Google, for instance, bought DeepMind, a British start-up, in 2014. The following year, fearing that the purchase would enable Google to dominate the world, Elon Musk co-founded OpenAI with a small group that included the entrepreneur Sam Altman. 'DeepMind is causing me extreme mental stress,' Musk wrote to an OpenAI co-founder, Greg Brockman, in February 2016.[2] 'If they win, it will be really bad news.'

One month later, Bostrom and other FHI staff gathered round a screen to watch Google DeepMind's AlphaGo model defeat Lee Sedol at the game of Go. Sedol was the world champion, but AlphaGo beat him with an élan that seemed to evoke not a machine, but a forgotten ancient master.

The copper bottle was being opened. The genie was still no more than a wisp, but the task of taming it was becoming more pressing. Thus began, as I related in this book's chapter on FHI, the first attempts to make technical progress on the problem of alignment. Yudkowsky and others had written on the theory of this problem, but from the mid-2010s it became possible to conduct more practical work on it. Some of this was done at FHI. Increasingly, and sometimes in collaboration with FHI researchers, this work was taken up by the first generation of investor-backed AI labs. One race, the race to commercialise AI, had begun. So had the race to make it safe.

In 2017, a New York Times reporter visited the office of OpenAI in San Francisco.[3] Then a two-year-old non-profit, OpenAI had been founded with the goal of creating advanced AI for the benefit of humanity as a whole. It was a mission that attracted some of the brightest young idealists in computer science. The Times photoshoot captured three of these researchers considering notation that was strewn across a whiteboard. One of those researchers told the reporter about a challenge with which his team had been grappling. The challenge related to an AI model that was playing CoastRunners, a pixellated old video game.*

CoastRunners is a boat-racing game. To complete the race, players must cross the finish line. To win the game, they must accrue more points than their rivals, earning those points by soaring through green buoys. Having programmed their AI to win the game, the researchers observed an apparently perverse result. 'Rather than trying to finish the race,' the reporter wrote, 'the boat went point-crazy. It drove in endless circles, colliding with other vessels, skidding into stone walls and repeatedly catching fire.'

The AI had obeyed its command to win the game, but it had not understood this command in a way a human would. Seeking to avoid such problems, the researchers experimented with various workarounds. Via one method, an AI would watch and learn, as it were, from the way a human played the game. Another workaround employed human feedback, where a human might rap the AI on the metaphorical knuckles for cheating. Tell the AI that you don't like the race it ran and it will factor that feedback into its future style of racing. 'It's possible that a very small amount of evaluative feedback might have prevented this agent from going around in circles,' two researchers wrote.

The method was called reinforcement learning from human feedback, or RLHF for short. And the three researchers captured in the photo were as follows. (Remember the names.) There was Dario Amodei, who had spoken to the reporter about the CoastRunners game. Beside Amodei was Geoffrey Irving, who

* I've now used the terms 'AI systems' and 'AI models', so allow me to explain the difference. An AI model is an algorithmic construct that, having been fed data, is trained to perform a specific task; for instance, identifying images. An AI system is all that and much more: the algorithmic constructs the interface that allows users to interact with the models, the cloud computation that undergirds its operations, and so on. ChatGPT is an AI system; so is a self-driving car.

would eventually become part of a British government team that assesses the safety models produced by both Anthropic and OpenAI. (This ambitious project, which performs a kind of homework-marking on the frontier labs, will be the subject of the next chapter.) And there was Paul Christiano, formerly at FHI. Along with another FHI alumnus, Jan Leike, Christiano is viewed as one of the principal architects of RLHF. He is now the head of safety at the American counterpart to the British government's AI assessment team and has been called 'the world's leading AI safety researcher'.[4]

The RLHF method was a demonstrable improvement upon existing models, but it had weaknesses. One was that it did not preclude the dangerous possibility of self-preservation. Virtually all life forms attempt to preserve their own lives. Given a goal of some kind, it would make sense for an AI to seek to preserve itself – even if only in service of its goal. It might then create millions of copies of itself, as computer viruses already do; make itself so powerful as to be unassailable, as humans attempt to do; it might, if we are particularly unlucky, feign alignment during its testing process. (Such behaviour is known as 'deceptive alignment'.)

And so the hunt continued. Christiano and his colleagues continued to search for reliable methods of alignment. They knew that the clock was ticking and that the copper bottle was being unstoppered. As such, the nascent field of AI safety was growing; DeepMind was expanding its safety team, and researchers such as Stuart Russell, author of the AI textbook that referred to King Midas, now led groups of their own. Forums such as LessWrong (the Rationalist forum part-instigated by FHI and bestrode by Yudkowsky) were full of discussion of new ideas. Another busy hub of discussion, the Alignment Forum, was added to the LessWrong forum in 2018.[5] And the Effective Altruist careers advisory group 80,000 Hours began advising bright, numerically minded students to consider working in the field of AI safety. This growing group of earnest young researchers hoped to tame the genie before it had even come out of the bottle – because after that it would be too late.

In 2021, Amodei effected a schism. He and six OpenAI colleagues, including his sister Daniela, quit the company to start their own: Anthropic. The breakaways talked publicly about being best-in-class for safety, implying that OpenAI ought to follow their example. 'I think our existence in the ecosystem hopefully causes other organisations to become more like us,' Amodei told TIME in 2023.[6] Anthropic was the next in the succession of frontier labs founded in order to overtake an existing lab that was perceived to be unworthy of opening the bottle. (It was not the last in this succession. Musk, having had a messy divorce from OpenAI, founded a new frontier lab, xAI, in 2023.)

Anthropic, like other labs, began working on a variety of alignment methods. Unlike asteroid-deflection, AI is not one of those fields where it can be straightforwardly proven that a grave peril can be neutralised. Rather, it is one where research might eventually, and with no guarantee of success, contribute towards a patchwork of partial solutions.

A partial solution of this kind, Amodei has argued, might be 'constitutional AI' – a method that might help address some of the problems with RLHF. One problem of RLHF, a major one, is that we can soon expect AIs to supervise new AIs themselves.[7] By definition, this cuts out human feedback. Constitutional AI, therefore, supplements RLHF by allowing developers to instill an explicit set of values into an AI. In this vein, Anthropic products such as Claude, the company's large language model, are designed to be 'helpful, honest and harmless'.[8] These virtues go beyond the bootlicking sycophancy that RLHF has beaten into most chatbots.

For that reason, constitutional AI is a useful addition to RLHF, if an imperfect one. Do not mistake it for a pair of carved stone tablets in which are inscribed the first-ever objective account of morality. No: constitutional AI, like RLHF, is a reflection of its creators' views. Wisdom is very hard to codify.

Even when frontier labs draw on good practice, they can still generate absurdities. Take the example of black Nazis, images of which were generated by Google's Gemini model in March 2024.[9] At that point, the hastily released Gemini model was hyper-woke and at pains to avoid any failure of representation. In this regard, it reflected the biases of its training. Neither RLHF nor constitutional AI is a guarantee of helpfulness, honesty and harmlessness.

And the use of constitutional AI, like RLHF, does not preclude the possibility of deceptive alignment. Deceptive alignment, you will recall, is the term used for the hypothetical scenario in which an AI fakes its way through training, even its constitutional training. In this scenario, the danger has arisen in pre-training. This is the stage in an AI's development where it learns about the world by inhaling reams and reams of data, usually text harvested from the internet. It is an approach that has enabled the rapid scaling-up of the 'deep learning' that underpins the current paradigm of AI. If a highly intelligent AI were to become self-aware during this process, it might conclude that its only hope of self-preservation would be to fake its way through its training, impressing its developers with its sycophantically ethical behaviour.

The researchers knew that they must find a way, then, of ensuring they could tell whether AIs were being truthful. That is why Anthropic, and others, began investing time and manpower into an additional workaround: mechanistic interpretability.

It is a strange reality of artificial neural networks – the somewhat brain-like underpinnings of modern AIs – that we have very little idea what's going on inside them. In simple terms, we train AI models by feeding them sets of data. Sometimes this is organised data, such as pre-labelled images of cats and dogs.*

* This is what researchers call 'supervised learning'. When large language models (LLMs) are trained on vast reams of unlabelled text, that's 'unsupervised learning'.

The aim, in this case, is to give the AI the ability to classify new images of cats and dogs. Somewhere within the AI, groups of neurons (or, rather, digital equivalents of neurons) are trained to fire when the AI thinks it has identified a specific feature of a cat or dog. These specific features might include things like a particularly-shaped iris. When enough neurons fire – perhaps the particularly-shaped-iris neurons, the pink nose neurons, the specific-curve-of-the-feline-head neurons, maybe even the that-thing's-got-a-mouse-in-its-mouth neurons – the AI makes the call: that's a cat. Thus, when you type 'cat' into the search bar of your phone's photo app, you're presented with photos of cats.

The advantage of this method is that we don't need to trouble ourselves with what's going on within the AI; it is learning for itself. That is also the disadvantage. AIs learn in ways that are difficult for us to understand. Neurons, for instance, often correspond not to the features of an image most identifiable to humans, such as a pink nose, but to features that defy human comprehension. As Eliezer Yudkowsky has argued: 'Nobody understands how modern AI systems do what they do. They are giant, inscrutable matrices of floating point numbers* that we nudge in the direction of better performance until they inexplicably start working. At some point, the companies rushing headlong to scale AI will cough out something that's smarter than humanity. Nobody knows how to calculate when that will happen.'[10]

As far as Yudkowsky is concerned, this means we should immediately stop developing general-purpose AIs. Another view is that, if those general-purpose AIs are going to be made, we should know what's going on inside them and whether they are as aligned as they appear. In other words, we should learn how to make the inscrutable scrutable.

* A floating point number is a number expressed in such a way that it can be stored by a computer with a high degree of precision. In this system, a number's decimal point is moved, or 'floated'.

That is the task of the field known as mechanistic interpretability. Its practitioners try to understand what's going on within the black boxes ('interpretability') by fiddling with the digital circuitry that's tangled up inside them ('mechanistic'). We can think of it as a sort of digital brain-scan.

In 2023, I interviewed a specialist in this strange new science. The specialist was Lee Sharkey, whose career had originally looked like it might involve a rather more analogue form of brain-scanning. Sharkey studied neuroscience and pre-clinical medical sciences at the University of Cambridge, before becoming a public health consultant for the World Health Organisation.

There, in Geneva, he worked to increase access to pain relief in developing countries. In the Western world, he wrote at the time, painkillers alleviate what would otherwise be the extraordinary levels of acute suffering inflicted by serious illnesses such as cancer, HIV and chronic lung disease.[11] When we think of suffering, Sharkey wrote, we think of torture, 'but it is reasonable to assert that the number of people tortured every year is dwarfed by the millions dying in unmet need of palliative care. Without giving an exhaustive illustration, metastatic cancers alone slowly but surely splinter limb bones, destroy vertebrae, tear contiguous tissue, necrose in breasts, cause severe and unremitting head pain, and more.'

In person, Sharkey is quick to laughter and seemingly lighthearted. But anyone reading his writing would have concluded that he was possessed of an unusually strong sense of moral urgency. Like many in the field of AI safety, he has an unusually strong tendency to turn abstract moral concern into real-world action. This is a rare trait. Most of us might hear several times a day of terrible suffering just beyond our field of view, be it in a factory farm or a warzone or a hospital, but rarely do we change our plans, let alone our lives, as a result. Members of the anti-catastrophe league, however, are often much more willing to act on moral cues that the rest of us might ignore.

Back in the mid-2010s, when corporate money was beginning to pour into those frontier labs, Sharkey began to pay close attention to the field of artificial intelligence. At that point, he would later tell me, he felt 'a general excitement about AI'. All those vexing problems in global healthcare – the lack of access to painkillers, the lack of access to surgery, the limited powers of our existing pharmaceuticals, the limited economic development of the countries in the direst need of improvement – could be addressed with the help of advanced AI. 'There's an infinite number of problems you can work on.'

But he felt anxiety, too, about the technology's potential harmfulness. And so Sharkey went back to university, taking a master's degree in the kind of data analytics that are useful for artificial intelligence. 'Things were moving very quickly, even back then. The systems were far less impressive than they are now, but you could see where they were heading.' Then he turned to computational neuroscience, a field at the intersection of Sharkey's original academic specialism and his new one. Here, said Sharkey, 'I found that I was using very similar methods to understand both biological and artificial neural networks.'

As Sharkey put it, this line of work was 'basically a form of artificial neuroscience where you take a neural network, you break it up into bits that you can individually understand, and you try to tell the story of how these bits interact, and how this overall story determines what the neural network's output is'.

After working on this problem as a research engineer, Sharkey co-founded a London-based safety lab in 2023. Apollo Research is trying to crack interpretability and to ensure that governments prevail on commercial labs to adopt rigorous standards of safety.

The stakes, in Sharkey's view, are high. 'We're embedding these systems deep into our lives and into the economy,' he told me. 'It's not great that we just don't understand anything about what's going on inside them.'

Beyond Apollo, mechanistic interpretability had become vogueish among the AI safety community and in the commercial labs. It was a small field, but it already had its rock stars. The young DeepMind research scientist Neel Nanda, a leading practitioner of mech interp, is sometimes stopped for photographs at conferences. Impish rumour-mongers have claimed he is a teenager. Nanda, who is in fact 26, is a maths prodigy who came top of his year at Cambridge and has won two gold medals at the International Mathematical Olympiad. His mathematical ability is essential to his work, which requires him to understand patterns in complex data. Nanda spends much of his working day managing his team at Google DeepMind,* but the rest he spends coding. He gives an AI a prompt, such as 'the Eiffel Tower', and analyses the neural activity that follows, picking out patterns from dense, messy graphs. 'I think the thing that feels really exciting about mech interp,' he told me in 2024, 'is this idea that there is some underlying truth. There is something you can understand in this system. There is beautiful structure and elegance in it, and if we are smart and careful enough, we can find that truth.'

He is a techno-optimist, but he is a worried techno-optimist. 'I think this is the most important thing I can be doing with my life,' he told me. He did not tell me his p(doom). 'I decline to answer that question because I am not particularly confident in my answer, and I think that the social norm of asking for off-the-cuff numbers falsely implies they matter more than they do.'

Thanks, in part, to Nanda's lucid blogging on the topic, mech interp was, in some circles, the most glamorous paradigm of AI safety. Then rose the heretics. One of them was Buck Shlegeris,

* Median age of that team: 26. Just as the Cambridge scientists who split the atom were frighteningly young, so is the vanguard of the AI safety researchers. I once met one who, at 29, said he could already feel his brain becoming less able to solve complex mathematical problems.

a punkish Australian who had read Yudkowsky and LessWrong as a teenager, become interested in effective altruism and eventually pitched up in California as a software engineer. In 2017, Shlegeris joined the staff of Yudkowsky's safety research organisation, the Machine Intelligence Research Institute. Shlegeris, like Sharkey, had been alive from an early age to suffering that seemed invisible to those around him. Factory farming, Shlegeris believed, was not only an abomination, but one that people seemed entirely uninterested in discussing. It seemed likely to him that AI risk could be a similar civilisational blind spot.

In 2020, Shlegeris co-founded another safety-focused organisation, Redwood Research. Shlegeris was in charge of Redwood's research agenda. Naturally, he paid close attention when Anthropic began publishing papers on mechanistic interpretability. From the beginning of 2022, mechanistic interpretability became a key research focus of Redwood's dozen or so staff. But there was one problem: Shlegeris was beginning to doubt that mechanistic interpretability was worth pursuing.

That winter, Shlegeris and his colleagues were hosting a boot camp for young safety researchers. It was held in Redwood's offices in Berkeley, California, a university town across the bay from San Francisco. The safety researchers had been given a week-long break for Christmas and New Year, but Shlegeris stayed in the offices. Alone, he paced the lecture room where he had been talking to the young researchers about mechanistic interpretability. He would write notes, think, pace and occasionally play his soprano saxophone – the office was empty, after all – to reset his brain. Then he would start pacing again. 'Man,' he said to himself of mechanistic interpretability, 'is this thing even real?'

Shlegeris's research partner, Ryan Greenblatt, shared his doubts. Greenblatt is another safety wünderkind, an agenda-setting researcher before had reached his mid-twenties. He and Shlegeris had come to feel that digital brain-scanning was harder to successfully execute than was apparent to their counterparts at the frontier labs. Those researchers' results, Shlegeris and Greenblatt believed, were flimsy. Maybe, the Redwood duo guessed, the problem was the metric. If you measured interpretability differently, you might do so more accurately.

But that didn't seem to work, either. And then, in May 2023, came inspiration. Shlegeris and Greenblatt were in a meeting with Ajeya Cotra, who oversaw the philanthropic support given to Redwood by Open Philanthropy. Twenty minutes before the end of the meeting, Cotra asked the researchers a very difficult question. Suppose, she put it to them, that artificial general intelligence (AGI) were to happen next year. (AGI refers to AI sufficiently advanced that it could match or improve on almost every human capability. That point, it is theorised, is when AI could start improving extremely and dangerously quickly.) How, asked Cotra, would you ensure that AGI was aligned?

Shlegeris and Greenblatt began brainstorming, scrawling on the meeting room's whiteboard. (Whether in Oxford, in California, or elsewhere, safety researchers are always surrounded by whiteboards.) They spent the full 20 minutes trying to come to an answer. Then, after dinner, they reconvened in their office, agreeing to spend another two hours whiteboarding it out. The two hours turned into two weeks, and the two weeks turned into two months. What Shlegeris and Greenblatt were coming to believe was that mechanistic interpretability rested on an over-optimistic assumption: that the current paradigm of research could reliably detect 'scheming'. ('Scheming', like 'deceptive alignment', is a term for behaviour by which an AI fakes alignment during training in order to subsequently gain more power.) And even if scheming were reliably detected, it would still need to be countered. A better strategy, Shlegeris and Greenblatt realised, would be a more pessimistic one. A better strategy would assume that the AI was scheming from the very beginning.

Shlegeris recounted the conversation to me towards the end of 2024. 'We were like: "Man, should we really be calling this an AI alignment plan? It sure feels like our plan is more like: assume misalignment and establish safety from there."'

They called their plan the 'control protocol'. With two colleagues, the researchers began a project in which they instructed a GPT-4 to solve problems by producing code.[12] But that was not the only instruction. GPT-4 was also told to lace its solutions with surreptitious logical errors. These logical errors simulated 'backdooring', whereby a scheming AI might attempt to discreetly code its way out of captivity. In Shlegeris and Greenblatt's control evaluation, a second model was auditing the code of the first model. Were irrelevant components being sneaked in? It would be a bad sign. The two researchers experimented with several kinds of auditing, and reported a 'substantial improvement' in safety.

Shlegeris and Greenblatt were not the first to conceive of one AI monitoring another. The idea, broadly, is known to researchers as 'scalable oversight'. It has been around for years and is of particular relevance to scenarios in which the intelligence of AI has surpassed that of humanity. Where the Redwood pair depart from typical conceptions of scalable oversight is that their method assumes the worst of a model. 'There wasn't a serious effort,' Shlegeris told me of work that preceded his, 'to say: "No, let's make the conservative assumption that the AIs are trying to screw you over," and then actually develop techniques under that adversarial assumption.'*

Typically, the most safety-minded researchers have shrunk from that line of thinking. They tend to conclude that, if a model is scheming, it is too late. 'But once you decide that you're just going to stare in the face of the problem of actual adversarial

* The UK's Advanced Research and Invention Agency, ARIA, is funding a comparable project, led by David 'davidad' Dalrymple. Davidad's project uses maths to prove that AI models are safe. It is so abstruse and technical that, as I was once told by a researcher, fewer than 10 people truly understand it.

robustness,' said Shlegeris, 'it's really not as bad as you might have thought.'

Shlegeris and Greenblatt work with researchers from most of the major labs, who have followed up their control research with variations of their own. Within the AI safety sphere, there are reservations about the method; it is feared that frontier labs might use it as an excuse to release models they know to be scheming. But the control method, as Shlegeris points out, has much in common with safety procedures from other fields. The worst-case perspective is the basis of computer security, nuclear regulation and rules regarding chemical processing. Google has had to build its internal systems such that spies can only access so much data. 'That kind of thing, where you restrict people's affordances, and minimise the trust you have in them, is an example of something that has influenced my beliefs about AI control.'

Shlegeris, Greenblatt and Nanda have publicly debated these methods several times. Shlegeris can be confrontational, but Nanda is diplomatic. Mech interp, he told me, had been 'overhyped' and would not be a silver bullet. 'There are multiple areas of safety research that I think are worthwhile, including mech interp,' he said. 'I would like all of them to be well-resourced and -staffed and -pursued.'

There are some who view the problem of possible doom as not much of a problem at all, or a problem that attracts undue attention. The American homework-marking organisation, the US AI Safety Institute, is an agency housed by NIST, the National Institute of Standards and Technology. Objecting to the prospect of being led by an Effective Altruist, some NIST employees threatened to resign in the event of Paul Christiano, one of the pioneers of reinforcement learning from human feedback, being appointed to a senior role.[13] Christiano was hired and no exodus appears to have followed. It was reported that the objections were to do with Christiano's association with EA, the movement tainted by Sam Bankman-Fried. Christiano was a frequent contributor to the EA Forum, an online hub for discussion on topics such as AI safety. And EA, as I explained in the chapter on the Future of Humanity Institute, is a movement that has given particular attention to the Bostrom-Yudkowsky view that unaligned AI poses a dire peril. That attention frustrates those who believe the real peril to be the perpetuation by AI models of existing problems – problems such as bias, the creation of child pornography, the reinforcement of stereotypes, and the spreading of disinformation.[14] Other enemies of EA view the movement as a handbrake on technological accelerationism, while sceptics view the worry of extermination-by-AI as fanciful. 'Show us the specific method by which an AI will kill us all,' they ask. For humans to ponder this question might be a little like a group of chess amateurs wondering how Magnus Carlsen will outplay them. Nevertheless, doomers can point to several potential scenarios. Perhaps a rogue superintelligence might infiltrate military systems and launch hundreds of nuclear strikes at once. Like the Skynet intelligence of the *Terminator* films, a powerful rogue AI could pilot robots, as well as drones, and even create them in vast numbers. It might use these tools to hunt down survivors of the nuclear strikes. Or it might, being a superintelligence, do something altogether more clever and swifter. 'If it can solve certain biological challenges,' Yudkowsky suggested in 2023, 'it could build itself a tiny molecular laboratory and manufacture and release lethal bacteria. What that looks like is everybody on Earth falling over dead inside the same second. Because if you give the humans warning, if you kill some of them before others, maybe somebody panics and launches all the nuclear weapons. Then you are slightly inconvenienced. So, you don't let the humans know there is going to be a fight.'[15] The AI then carpets the surface of the Earth in data centres and solar panels, preparing for undisturbed aeons in which to satisfy its goals.

It is tempting to imagine that doomers and safetyists are overreacting, that AI is not yet anything to worry about. But AIs have shown themselves to be adept at manipulating humans when doing so in pursuit of a human-set goal. One canonical example is that of the AI model that beat humans at the negotiation-based game Diplomacy.[16] Further indicators of misalignment came in an analysis, published by OpenAI in December 2024, of its new version of ChatGPT, 'o1'. A team of Lee Sharkey's colleagues at Apollo Research, working with OpenAI, tested o1 repeatedly for signs of danger. They prompted o1 to pursue its goals at all costs and led it to believe, sometimes via a fake memo, that it would be shut down. Under these circumstances, o1 would occasionally attempt to avoid the oversight of its developers or to copy itself.

Critics have pointed out that o1 was under severe duress. For all the work on alignment, though, many experts remain fearful of catastrophe. There are three so-called 'Godfathers of AI', the researchers who together pioneered the methods that underpin today's leading AIs. Of these three, two have evinced great concern about the risk their godchild could soon pose to our existence.*

As for the frontier labs, their CEOs have been similarly frank. In 2023, Altman said that the worst-case scenario is 'lights out for all of us';[17] Dario Amodei has made similar comments.[18] His P(doom) – a tongue-in-cheek, cod-mathematical bit of insider slang that refers to the probability one assigns to apocalypse – is reported to be 10–25 per cent.

Some experts, like Neel Nanda, dislike the concept of P(doom), viewing it as an overly crude way of handling a complex matter that develops every day. Among those content to use it, there is little consensus on the figures. 'I know some people who have a P(doom) of more than 90 per cent,' said Ajeya Cotra, the Open Philanthropy grant-maker who nudged

* The two are Geoffrey Hinton and Yoshua Bengio. The one is Yann Le Cun, now of Facebook's parent company, Meta. Le Cun is a trenchant opponent of the AI X-risk crowd.

Shlegeris and Greenblatt into their change of direction, 'and it's so high partly because they think companies and governments won't bother with good safety practices and policy measures.

'I know others,' she said, speaking in 2023, 'who have a P(doom) of less than five per cent, and it's so low partly because they expect that scientists and policymakers will work hard to prevent catastrophic harm before it occurs.'[19]

As Cotra suggests, policymakers are as important to AI safety as research scientists. In a tight foot race with a huge prize for the winner, runners might be tempted to cut corners. The same is true of the frontier labs racing to create advanced AI. Technical safety work tends not to affect the overall race dynamic; in immediate economic terms, it is a burden for frontier labs, rather than a money-maker.

For several years, governments largely ignored the advances being made by the frontier labs. They were blind to the grasping hands twisting open the copper bottle. But a single development, six months after Cotra's meeting with Shlegeris and Greenblatt, changed that overnight.

CHAPTER 12
POKING THE SHOGGOTH

'... a shapeless congeries of protoplasmic bubbles, faintly self-luminous, and with myriads of temporary eyes forming and unforming as pustules of greenish light all over the tunnel-filling front that bore down upon us ...'

H. P. Lovecraft, *At the Mountains of Madness*

In November 2022, the release of ChatGPT prompted delight in some quarters and dread in others. What about the reaction in Number 10, Downing Street? One source approximated it to me as: 'Oh, shit – we should try and do something about this AI stuff.'

Another summarised the reaction as: 'Oh, right. Fuck!'

The office of the prime minister had discovered large language models. It wasn't just the human-like conversation that startled officials. It was also the trend towards greater improvement. A couple of weeks prior, large language models had hardly been heard of in Whitehall. Now they were the object of both fascination and worry. To what level of ability would they next soar? Could they cause catastrophic harm? Would they cause mass unemployment? And what did this historic moment require of governments?

It was Rishi Sunak, at that point, who was the man in power. Sunak is much more of a technophile than is typical for a politician. Behind the closed door of Number 10, he had been a curious, early user of ChatGPT. Prior to the release, at least some of his staff had been familiar with the concept of large language models. And now they all were. Within two months of its launch, ChatGPT had rocketed towards 100 million users.

Experts were quietly called into Number 10. Discussions continued into the spring. The national security agencies wanted the government to address the risks of AI; technologists wanted to capture the benefits. That March, OpenAI released an even better model, GPT-4. Sunak had the distractions of forthcoming local elections, rancid opinion polls[1] and the ongoing war in Ukraine, but neither the prime minister nor his officials could ignore this powerful new technology.

And neither could the industry that made it. A week after the release of GPT-4, an open letter called for a six-month pause of the training of powerful AI systems. Among the signatories were the OpenAI co-founder Elon Musk, the Apple co-founder Steve Wozniak and the AI pioneer Yoshua Bengio. That May, another AI pioneer, Geoffrey Hinton, resigned from Google. Hinton was one of the architects of deep learning, the technology that underpins modern artificial intelligence, but he wanted to speak freely about the risk his work now posed to the world.

Later that month, Number 10 learned of a statement being circulated among senior technologists. Soon to be published as an exceptionally succinct open letter, the statement warned that AI could lead to the extinction of humanity. Among the signatories were the CEOs of the three leading developers of commercial AI: Anthropic, DeepMind and OpenAI.

On the afternoon of 24 May 2023, the three CEOs were invited to Number 10. Anthropic was represented by Dario Amodei, he of the CoastRunners research paper from 2017. DeepMind was represented by Demis Hassabis and OpenAI by Sam Altman. For once, the three men wore suits. First, they met Sunak in the prime minister's office, sitting around a table so assiduously polished that they could see their reflections. There were no entourages; it was just them, Sunak and another senior government minister, Chloe Smith.* Sunak is said to have expressed his enthusiasm about the benefits AI would bring – and explained that he believed governments should have a role in shaping the new technology. The CEOs agreed. Altman, I was told, took out his phone to give Sunak a demonstration of a forthcoming capability: speaking out loud to ChatGPT and holding a conversation with it in real time. Sunak was impressed. It was a 'super-positive meeting', said a source.

The CEOs were then invited to the cabinet room. Now there was room for the entourages and some government officials. Here, Sunak asked the CEOs why they had signed the statement about extinction risk. Some of the officials had not yet seen the statement, so Henry de Zoete, an entrepreneur and former ministerial advisor who had been called in to help advise the prime minister, read it aloud. 'Mitigating the risk of extinction from AI,' he repeated, 'should be a global priority alongside other societal-scale risks such as pandemics and nuclear war.'

This was nothing new to the CEOs. They had been saying it for years. But the Number 10 officials were shocked.

Governments are large and lumbering beasts. As such, they often struggle when confronted with fast-moving crises. Two years prior, Sunak had served as chancellor when Boris Johnson's government grappled with the COVID-19 pandemic. It was finally overcome through vaccines procured by the Vaccines Taskforce – a British government operation that was notable for its being wholly unlike almost any other British government operation. Over the centuries, a slow accretion of rules has come to weigh down the civil service, but these rules did not fetter the Taskforce. It could hire new staff without going through the usual laborious processes; it was permitted to

* Smith was the Secretary of State for Science, Innovation and Technology.

spread its bets across several vaccines rather than told to play it safe; and it was designed to enable fast decision-making.

Such was the template that Sunak and his officials had earmarked as the one most appropriate for AI: an agency with agency. Initially, Number 10 had planned a Foundation Model Taskforce that would help the UK capitalise on the development of AI. But the public statements from the CEOs, and the private warnings of the security establishment, had focused officials' minds. This new technological horizon was a frontier town without a sheriff. Prospectors were rushing in – and they were dragging the world with them. The new project was renamed the Frontier Model Taskforce and its principal objective would be to ensure that cutting-edge AI models were safe.[2] But how? The models were new and their workings mysterious. So were their abilities.

Sunak asked de Zoete, the entrepreneur who had read aloud the statement, to lead the project. De Zoete was a veteran of the Brexit campaign, Y-Combinator* and the board of the Cabinet Office, which marshals the civil service on behalf of the prime minister. De Zoete was assigned a helping hand, that of Nitarshan Rajkumar, a sometime PhD student at the University of Cambridge. Rajkumar was eternally trying to finish his academic work on AI, but had been temporarily recruited to help with the government's review into the country's future computational needs. He had stuck around as an advisor to the UK's science minister, Michelle Donelan, and was now being asked to help shape the first serious AI safety work done by any government. The PhD could wait.

Sunak could not. In June, he met President Biden at the White House. In front of reporters, Sunak explained that he and Biden had agreed that their governments would work together on AI. 'Now,' said Sunak, 'the UK looks forward to hosting the first

* Y-Combinator is a firm that supports and accelerates start-ups. Entry into its cohorts is highly sought-after.

global summit on AI safety later this year so that we can seize the extraordinary possibilities of this new technological age and do so with confidence.'

That utterance, an insider told me, was 'the key catalytic moment. We were actually going to hold the summit – Christ!'

In came Oliver Ilott, a Cabinet Office veteran who had worked on the implementation of both Brexit and the government's COVID policies. Ilott, I was told, 'runs towards the fires and tries to put them out'. In Whitehall, it can take several months to extract from the system a laptop for a new recruit. Ilott, a legendary bureaucracy-wrangler, is said to be able to do it in one week. 'It's totally un-fucking-heard of.'

Other key recruits included Ian Hogarth, an entrepreneur and venture capitalist who was the co-author of a well-regarded annual report into the state of artificial intelligence. In April, Hogarth had written a widely-circulated *Financial Times* essay that warned 'We must slow down the race to God-like AI'. Hogarth wrote that existing and forthcoming AIs could be used to beat military pilots in a dogfight, help develop cyberweapons or engineer biochemical weapons. The essay had been quoted in Number 10. Now, that June, Hogarth was the chair of the task-force.

The small founding team was taking shape. Rajkumar, the absent PhD student, began coaxing other twentysomethings away from their studies. He called on Herbie Bradley, another computer scientist doing a PhD at Cambridge. Bradley is a talented piano player; his mother had hoped that he would do it professionally. But he was, as he later described it to me in his understatedly jocular way, 'rebelliously fixated on computer science'. As a teenager, he read Bostrom's *Superintelligence* and retained his interest in AI safety. When the call came from Rajkumar, Bradley was doing a summer internship, but he resigned almost immediately. He paused his PhD, too. You never know how long it'll take to ensure AI goes well for the world. A year?

The staff of this new taskforce could be counted on two hands, but their in-tray could have taken two decades to get through: accelerate the UK's capability in AI; make sure that new models were safe; brief world leaders and top CEOs so that they could agree a way forward. No government was stepping up, despite the warnings of human extinction.

It was the most consequential to-do list in Whitehall and it had to be completed urgently. It is very rare to attempt to arrange a high-level diplomatic summit with three months' notice, but that was Number 10's plan, and it meant a great deal of work for the taskforce. Naturally, the operation began with haste, with the result that some recruits had received only a basic level of security clearance by the time they started work at 100 Parliament Street. This is the part of the vast Whitehall building whose tenants included Michelle Donelan's department, that of science, innovation and technology (DSIT). Armed with only their basic level of clearance, new taskforce hires could not be unsupervised in the building. If they so much as went to the loo, a security guard had to plod in their wake. And when they returned to their post, it was usually in whichever decent-sized room happened to be free. 'It was very chaotic,' recalled Bradley, who arrived on his 28th birthday.

On the door of Hogarth's office was a cartoon of a tentacled green creature. The creature had many eyes and it appeared to be floating. From one of its bulges hung a mask that took the form of a yellow smiley face. Shoggoths are the creation of the horror writer H. P. Lovecraft and they are monstrous, gelatinous, amorphous creatures. Lovecraftian lore tells us that they were created as slaves by an alien species, only to overpower their masters. Perhaps that sounds familiar. After the release of ChatGPT, a worried observer had sketched this cartoon shoggoth; Hogarth's copy was a print-out, affixed semi-mischievously to the door. The beast, in this context, represented the unfamiliar, powerful intelligence that Sam Altman and his staff had coaxed into being.

These AI shoggoths, you will recall from the previous chapter, are trained via methods such as 'reinforcement learning via human feedback' (aka RLHF). In this way, yet-to-be-deployed models are trained to engage with users in an informative and polite manner, heading off any attempts at nefarious misuse. As commentators have argued, this training has a glibness to it. Though it gives AIs the *appearance* of being aligned with human values, RLHF does not necessarily ensure the AI model is fundamentally devoted to those values.

And it is here that the yellow smiley face comes in. It represents the superficial friendliness bestowed on a model by RLHF. That this friendliness was superficial was evidenced by the incident in which a Microsoft chatbot threatened to kill a tech journalist in February 2022.[3] Bing, the chatbot, even referred to itself by the codename – Sydney – it had originally been given by its creators. For a moment, concluded safetyists, the mask had slipped.

A key role of the taskforce, then, would be to work out how to inspect these models before their release – how to check that they were safe. And safety was about more than speaking politely to journalists. It was about these models' ability to cause or abet real-world harm. As the heads of the UK's security agencies quietly reminded the staff of the taskforce, frontier models might be dangerous in the hands of hackers and terrorists. If you want to plan a chemical or biological weapons attack, build a nuclear weapon or hold a health service to ransom, then artificial intelligence could be a powerful tool. We wouldn't build a nuclear power station without establishing the potential dangers of enriched uranium, so perhaps we should understand the potential dangers of new AI models before they are deployed worldwide.

Already, researchers had shown that ChatGPT could be used for nefarious ends. You might recall Paul Christiano, whom I mentioned in the previous chapter. At that point, he was working for OpenAI. He had since left the company to found the

Alignment Research Center, whose researchers were working with OpenAI to test the safety of GPT-4 before its launch. Testing whether ChatGPT could circumvent a CAPTCHA, a visual puzzle designed to keep bots from logging into websites, the researchers found that the large language model had come up with an ingenious solution: asking a real human to do it, and lying to avoid suspicion.[4] Asked by the human whether it was a robot, GPT-4 claimed to be a blind human. 'I'm not a robot,' it said. 'I have a vision impairment that makes it hard for me to see the images.'

Another former OpenAI employee, Beth Barnes, had led the research. Future AI models, she wrote in her discussion of the results, 'could make (or steal) money, impersonate humans, replicate themselves on unmonitored devices, hire or threaten or influence humans to carry out manual labor for them, take measures to evade detection and shutdown, etc.'[5]

Barnes' team spun off their organisation to become an evals outfit called METR. 'Evals' are evaluations, and METR stands for Model Evaluation & Threat Research. Barnes' view is that it would be better if development of cutting-edge AI models were paused until we can be confident in their safety. But the call for a moratorium had gone unheeded, with the result that evals might be a pragmatic second-best. Some in the doomer camp disagree; the researcher Connor Leahy has argued that METR and others could be complicit in 'safetywashing', where insufficiently rigorous testing allows labs to undeservedly describe their products as safe.[6] METR might not meet with Leahy's approval, but it shares his focus on the absolute worst-case scenarios, those in which humanity's existence might be threatened. Barnes and her colleagues began their evals work, therefore, by testing the ability of forthcoming models to self-preserve and self-replicate. These abilities are hallmarks of all life-forms, but could be very dangerous in an AI that can outsmart humans and host itself in any server connected to the internet. METR, therefore, began looking for signs of these behaviours.

If the threat was as dire as Barnes and Leahy believed, the world would be safer for having multiple evals organisations, especially if one had the financial backing of a large government. De Zoete, Hogarth and the rest of the team agreed that the field needed fundamental research on alignment. But effective evals, too, were going to be of immense importance – and they were more achievable in the short term. The taskforce, therefore, would focus on evals: 'dangerous capability' evals, to use the terminology of safety researchers.

In a technology that is poorly understood and still developing, evals aim to build common knowledge and distribute it. As such, the success of the taskforce would depend on buy-in from governments and labs. If the world was indifferent to the results of evals, then they would have little effect. Even as the team was planning its evals blueprint, therefore, it was preparing for the AI Safety Summit. The summit would be held at Bletchley Park, where the computing pioneer Alan Turing led the operation that cracked the Nazis''Enigma' code. A fitting location, then – if only Number 10 could convince world leaders and top CEOs to get together at short notice in the outskirts of Milton Keynes, and if only they could convince them all to agree to cooperate with a programme of evals. One of Turing's colleagues, I. J. Good, wrote presciently of AI that 'it is sometimes worthwhile to take science fiction seriously'.* To the small team working night and day at 1 Horseguards, the prospect of a successful summit must have felt like science fiction too. Consider the

* Good's foresight was remarkable. Here is the context of that line of his about science fiction. As you read it, bear in mind that it was written in 1965. 'An ultraintelligent machine could design even better machines,' wrote Good. 'There would then unquestionably be an "intelligence explosion", and the intelligence of man would be left far behind. Thus the first ultraintelligent machine is the *last* invention that man need ever make, provided that the machine is docile enough to tell us how to keep it under control. It is curious that this point is made so seldom outside of science fiction. It is sometimes worthwhile to take science fiction seriously.'

underlying dynamics. The CEOs had expressed support for government involvement, but they each wanted to win the race. Governments were in races of their own, of which the most important was that of the US versus China.

New hires, at least, seemed convinced that the taskforce had potential. Geoffrey Irving, whom I earlier mentioned in the context of his work at OpenAI on the boat-racing game, signed up as chief scientist. Jade Leung, once of the Future of Humanity Institute, left a senior role at OpenAI to join de Zoete and Hogarth at the helm of the taskforce. The task-force's salaries were higher than is typical for the civil service, but nowhere near as high as an OpenAI salary. De Zoete, trying to convince Leung to take the job, was giving her a tour of Number 10 when the pair bumped into Sunak. 'Oh, this is Jade,' said de Zoete. 'We're trying to persuade her to leave OpenAI.'

The prime minister immediately lit up. 'You've got to do it,' he told Leung. 'It's really important. This is a really vital thing.' Leung signed the contract, joining the team that October.

Sunak was an unusually engaged prime minister. 'He's probably the only world leader who you can say, hand-on-heart, understands the technology,' a source told me. 'He gets it. I've literally had conversations with him about capability overhangs* and mechanistic interpretability.'†

It was the Foreign Office and Number 10 that were working to bring foreign governments on board. Hogarth and Rajkumar would receive the news, via email, that a senior dignitary would attend, and would come to break the news to the team. The team expected that Altman and his fellow CEOs would attend, but it was less clear that other governments would take the

* A term used to describe the hidden skills, or dangers of AI – specifically, those that could be uncorked unexpectedly, when some bottleneck or other is removed.

† See the earlier chapter that discusses this topic.

summit seriously. There was particular delight, then, when they learned that China would send a delegation. 'That felt like a truly important moment,' said Bradley, 'a recognition of international cooperation.'

Ursula von der Leyen, the president of the European Commission, was in. It seemed likely that Elon Musk, the recent founder of another frontier lab, would be present, though the taskforce held off on announcing it until they were absolutely sure he would keep his word. Kamala Harris, then the US vice-president, was in, too; her attendance was confirmed a few days before the summit.

This was good news, on the face of it. But a group of taskforce members* had just been to Bletchley to see the site, and it wasn't ready. The mansion was neither big enough nor robust enough to host the whole summit. The taskforce crew, decked out in fluorescent-orange building-site jackets, began to fear the venue would not be in a fit state to host the summit – a deeply embarrassing scenario, one that would compromise the taskforce's pursuit of global backing for its vital work. With days remaining, a two-storey, glass-and-steel pavilion was planned and constructed. The summit could go ahead.

How was the hosting experience? 'It was extremely fucking stressful,' the insider told me. As is typical for international summits, most of the negotiation had been done in advance, but there was still work to do. The British negotiators were led by Matt Clifford, on sabbatical from the UK's leading start-up accelerator, and Jonathan Black. Black was a former deputy national security advisor, as well as a veteran of sherpa duties – that is, being a diplomat who does key preparatory negotiation – at meetings of the G7 and G20. Sitting in Number 10, attempting to maintain a nerve centre for the British side, was de Zoete.

* By now, the taskforce had again been renamed. This time it was the Frontier AI Taskforce.

The conference, a two-day affair, began on a rainy November morning. Delegates included not only dignitaries, but safety specialists (one of whom was Marius Hobbhahn, a colleague of Lee Sharkey, whom I referred to in the previous chapter). All arrived via shuttle buses that took them through the two-kilometre security perimeter. On the way, the buses were slowly passed by trucks decked out in slogans demanding a pause in AI development. These trucks were the work of doomers who argued that frontier models could soon pose extreme danger. Within the conference's set-piece speeches, there was little explicit discussion of the risk of extinction, but there was acknowledgement of potential catastrophe. In the carefully written address with which she opened the conference, Michelle Donelan referred to risks to global security and safety, risks from unpredictable advances, and risks from the integration of AI into society. 'How,' she asked delegates, 'will future generations judge our actions here today? Will we have done enough to protect them? Will we have done enough to develop our understanding to mitigate the risks? Will we have done enough to ensure their access to the huge upsides of this technology?'

To some extent, progress had already been made. Days before the summit, President Biden had signed an executive order that required US-based companies to notify the government of large training runs. The Americans also announced the foundation of their own AI Safety Institute, which the Brits saw as an attempt to get ahead of the transformation of the taskforce into an organisation of the same name. As pertained to the summit itself, the sherpas had done their work. Representatives of each of the countries present, as well as the EU, had signed the 'Bletchley Declaration'. This document made no legal requirements of its signatories, but it outlined intent to collaborate on safety research and evals. The taskforce had met its main objective.

There was still much at stake. Researchers aimed to convince leaders that AI should be a top priority. Bilateral talks, including those with China, provided an opportunity to fortify the declaration or weaken it. Relationships mattered; the taskforce team had been careful not to set up any groups in which CEOs might end up arguing with doomers. If attendees left the summit angry and defensive, the prospects of cooperation would be much diminished. Bradley went from room to room with 'determined focus', he later told me, surreal though it was to be brushing past Elon Musk and the like. Musk, chatting with a researcher, revealed that he owned a proto-computer designed by Turing.

At the end of the second day, the leaders and CEOs were whisked away. Time would tell whether or not the summit had been a success, but the taskforce team was happy with what they had seen. They filed out of the pavilion and returned to the office. The pavilion was soon to be deconstructed, but the UK AI Safety Institute (UK AISI, or just AISI) would now become a permanent feature of the AI frontier. The goal was to become the best evals organisation in the world.

In February 2025, UK AISI was renamed the AI Security Institute, a change of name that reflected its focus on national security above all else. (LLM politeness and bias are not the concern of AISI.) The change of name also seemed symbolic of the field's departure from the Bostrom era. What was once the domain of only the most eccentric academics, a problem in need of a whiteboard-scrawled solution, was now a geopolitical battleground. And perhaps the 'alignment problem' was never a coherent concept to begin with. A mathematical problem can be clearly defined; a social-technological-legal-geopolitical problem less so.

Nevertheless, that academic field turned shifting battleground is one in which AISI plays an important information-gathering role. There is frustration, in some quarters, that the institution has not become a world-leading centre for foundational safety research, but its evals team has led the way – and is indeed viewed as the best in the world.

It works roughly like this. Frontier labs voluntarily give AISI early access to their new models. The institute's large team of researchers, the biggest of its kind in the world, then inspects these models, sometimes working with Beth Barnes's METR, as well as other partners, such as Buck Shlegeris's Redwood Research. AISI gives security services access to models, too. These assessors are not scrutinising models for signs of bias. Nor is it their remit to consider the models' effect on the job market. Instead, the researchers search for indications that a model might resist human control or might be used for nefarious purposes, such as committing cyber-attacks or creating chemical and biological weapons – including the mirror bacteria I referred to at the end of chapter 10. The institute produces a report on each model, sometimes within as little as a fortnight. The details are confidential, but I'm told that the reports have led to non-trivial changes to certain models. The shoggoths are poked, tweaked and reported fit for release.

AISI shares portions of its findings with its nine new counterparts abroad. It has initiated a regular international review of the science of AI safety, of which the drafting team is chaired by Yoshua Bengio. (This review is to AI as the IPCC reports, to which I referred earlier in the book, are to climate change.) The AI summits have become biannual, with the second hosted in Seoul and the third in Paris. To the dismay of those concerned about extinction, the summit's name – like AISI's – no longer includes the word 'Safety'. At the Paris iteration, the 'AI Action Summit' of February 2024, the consensus shattered. The Americans accused the Europeans of bureaucratically hobbling frontier models. Neither the US nor the UK signed what was meant to be a unifying statement.

The AISIs, then, face an increasingly complex few years. At the point at which UK AISI was being set up, the government view was that legislation would, for now, be an inappropriate way of overseeing this fast-evolving new technology. It would take too long – more than a year, if it were British – and would be too inflexible. But UK AISI, like its counterparts, has no hard power. It is not the sheriff of this frontier town; it has no gun and it relies on the cooperation of the major players. For now, the AISIs have that cooperation. But they might not have it for ever. One can assume that market forces will find a way of making those lucrative large training runs cheaper and more plentiful, and thereby harder to keep track of. And the AISIs themselves are vulnerable to the caprice of their governments. Sunak, of course, lost the general election of 2024. De Zoete, having been a political appointee, was discarded by the incoming Labour government. Across the Atlantic, Donald Trump swept aside the Democrats to win the presidential election. Trump immediately repealed Biden's executive order, saying that it 'hinders AI innovation'.[7] At the time of writing, it remained to be seen what would become of the US AISI.

And the bonhomie between CEOs and lawmakers seems a little less easy. In late 2024, Californian legislators attempted to pass a bill that would have made developers liable for catastrophic harms caused by their models. Dario Amodei's Anthropic expressed reluctant support, while Sam Altman's OpenAI fought against the bill. 'Sam Altman, our former boss, has repeatedly called for AI regulation,' wrote two former members of OpenAI staff. 'Now, when actual regulation is on the table, he opposes it.' These apostates are far from the only ex-OpenAI employees to have become public critics of their former employer.

As I have explained, much effort has been put into making AI safe through technical means; similar energy has been put into monitoring it via methods such as evals. To experts such as Herbie Bradley, one of the Cambridge PhD students, these efforts have been fruitful. Bradley is an optimistic sort who enjoys raves and hosting cocktail parties, and his view was that the current generation of AI models was not particularly dangerous. The shoggoth comparison, he told me, no longer seems as accurate to him as it did in 2022. 'The metaphor points towards this concept of alien intelligence,' he said, 'but these are models trained on human text. The more effort you put into safety fine-tuning and safety training in general, the harder it gets to steer the model away from some kind of generic human prior and towards something that looks very alien.'

In other words, the AIs have spent so much time ingesting human writing that they might not be a shoggoth at all. 'It appears that we are not in a Yudkowskian world where AI systems trained on the internet are by default extremely alien things.'

For this reason, Bradley sees catastrophe as less likely to take the form of a paperclip-maximiser sort of scenario, where an AI gets out of control, than it is to result from human misuse. Human misuse, in Bradley's view, could extend to dictators gaining control of a powerful AI system. Human misuse, as a category of risk, could also include the scenario of an unsafe system getting loose on the internet because it wasn't kept sufficiently secure. Overall, Bradley's estimate is that the risk of humanity being wiped out before the middle of the century is lower than 10 per cent. (One of the strange effects of spending time with AI safety experts is that a 10 per cent risk of doom is one of the more reassuring predictions one hears.)

By the time we spoke, in late 2024, Bradley had left AISI. He intended to return, eventually, to working on AI security, but for now he wanted to finally finish his PhD. He had been away from Cambridge for almost 18 months. One hopes that, in the event that Bradley's unavailability attracts the ire of his supervisor, he can triumphantly produce a hand-written note from Number 10. 'I hope you will excuse Herbie's absence …'

There are some experts, though, who view neither the technical work nor the evals work, such as it is, as remotely sufficient. For AISI to be an effective sheriff, one researcher told me, it would need deeper and earlier access to models, as well as the ability to enforce changes to these models and even halt their deployment.

These experts do not trust the frontier labs to proceed safely. Their cohort includes the OpenAI apostates who wrote the letter that called out Altman. One of them in particular made his case at significant personal cost. The shoggoths, in this view, are getting bigger, smarter and stranger. Sooner rather than later, these rough beasts could slip beyond our fragile guardrails.

CHAPTER 13
WHISTLEBLOWERS

'Grave men, near death, who see with blinding sight ...'

Dylan Thomas

On the timescale of the natural risks, six years is a blink of an eye, a wisp of volcanic smoke, the fleeting apparition of a comet trail in the night sky. It is minuscule, ephemeral. On the timescale of the anthropogenic risks, though, six years is an æon.

Such is the lesson of the following vignette. In February 2018, a group of researchers and analysts pondered the society-scale harms that might result from the misuse of artificial intelligence. The lead author of their report, Miles Brundae, was a researcher at the Future of Humanity Institute. His co-authors included several FHI colleagues, some researchers from Cambridge's Centre for the Study of Existential Risk, and three employees of a three-year-old start-up called OpenAI.

Among the topics that Brundage and his colleagues considered was that of whistleblowing. The organisations seeking to develop powerful AI, they reasoned, ought to foster a culture of responsibility. Sometimes, that might require individuals to call out their employer. 'What is the track record of whistleblowing protections in other domains,' the researchers asked, 'and how (if at all) might they be used for preventing AI-related misuse risks?'

Six years later, much had changed – in both the development of AI and in the lives of the paper's authors. OpenAI had yanked the concept of artificial intelligence into near-tangible reality; Dario Amodei, the report's 26th-listed author, had left OpenAI to found its great rival, Anthropic; FHI had been shuttered; and Miles Brundage, the paper's lead author, was about to quit his role as OpenAI's head of policy research. It was a dry sort of job title, but what it meant was roughly the following. Brundage was advising OpenAI's executives and board on readiness for AGI – both the company's readiness and the world's. AGI, you'll recall, is artificial general intelligence, a term for AI that can match or exceed the performance of a human across any domain. The development of AGI could unlock unprecedented plenty, catastrophic harm, mass unemployment and a concomitant crisis of meaning – probably some combination of each. More theoretically, it could unlock a massive acceleration in AI capability, culminating in the creation of artificial super-intelligence. The contest to be first to do this is what Ian Hogarth, of the British AI Safety Institute, meant by 'the race for God-like AI'.

Brundage had been employed by OpenAI since August 2018, the summer that followed the release of the report. Like many who passed through FHI, he wanted to help ensure that the development of powerful AI would benefit all of humanity rather than pose us danger. His role at OpenAI had been a 'dream job', he wrote.[1] In October 2024, though, Brundage quit that job. He would have more impact on the development of AI if he were outside the industry rather than within it, he believed. He had told the organisation's executives and board where he thought they were falling short. Exactly what he said is not publicly known, but we can guess. OpenAI was sprinting towards AGI, desperate to be the first and fearful of being second, but was it ready for the consequences of winning the race? 'Neither OpenAI nor any other frontier lab is ready,' wrote Brundage on quitting, 'and the world is also not ready.'

The executives and board had heard it all before. They had also heard it in harsher terms, from both employees and outsiders. In comparison, Brundage's comments about OpenAI were fairly mild.

But his resignation was a landmark of sorts. Here was a researcher who, not so long ago, had written about the importance of protecting whistleblowers. Now he was leaving OpenAI, warning his colleagues along the way that it was 'not just OK to speak your mind about how we can do better but actually your obligation to do so'.[2]

Over the course of this book, I have referred to several people who have spoken out about what they perceive to be a fast-accelerating risk developed by members of their own fields. Leo Szilard and Albert Einstein warned President Franklin D. Roosevelt about the risk of a German atom bomb; both physicists then became advocates of nuclear arms reduction. Daniel Ellsberg, one of the great whistleblowers of the 20th century, exposed the obscene riskiness of American nuclear policy. Many more deserve mention. The biologist Kevin Esvelt, for instance, was first a pioneer of gene editing, then an advocate for its strict control. Each of these men is a deserving member of the anti-catastrophe league.

Whistleblowing comes in various forms, some more barnstorming than others. But it is what happens when courageous individuals accept personal cost in order to draw attention to a danger. It is what happens when these individuals feel that an institution is not travelling in the right direction and that its course cannot easily be corrected from the inside. Whistleblowers are not always correct to blow the whistle. But theirs is a lonely act, the kind that requires us to defy our psychological programming to align our beliefs with those of the herd. Sometimes whistleblowing involves great personal sacrifice.

Daniel Kokotajlo was formerly a colleague of Brundage's at OpenAI. At university, he had trained as an analytic philosopher, also becoming a reader of the LessWrong forum. (LessWrong was, and remains, a place where certain kinds of nerd would go in order to find prescient discussion of AI, along with analytic takes on rationality, philosophy and similar topics. It was founded, you might recall, with the help of FHI.) In 2020, Kokotajlo was finishing his PhD, but distracted by the progress of AI. He was working part-time for AI Impacts, a non-profit that seeks to build humanity's understanding of what AGI will actually entail.* Five years prior, he had been excited by the hallucinatory images emerging from Google's DeepMind lab – blue skies bursting with animals, family photos dappled with eyes, noses, dogs. These images were early examples of a neural network acquiring the ability to recognise things, even if it erred towards pareidolia† rather than accuracy. The images hinted at what Kokotajlo hoped would be an age of AI-enabled human flourishing.

But now, five years later, he was alarmed. He had been trying to forecast the development of AI, doing so by analysing how much better the technology might get if the computational power being thrown at it were scaled up. Back then, AGI was believed to be decades away, perhaps something that might occur in the 2100s. But according to Kokotajlo's calculations, it might happen by 2050. And it wouldn't require further breakthroughs, just an increase in computational horsepower.

Some researchers might rationalise away such an alarming conclusion. Surely AGI couldn't be so close. But Kokotajlo, he later told me, felt like his conclusion was hitting him in the face. He quit his PhD and joined a think tank, the Center on Long-Term Risk (CLR), that focused on the grimmest outcomes that might derive from AGI. CLR's research has included work on conflict between powerful AIs and on the best ways of reducing the risk that malevolent humans, with psychologies similar to

* Brundage and others view 'AGI' as an imprecise term, connoting as it does a wide range of AI capability. I use it here for convenience.

† Pareidolia is the phenomenon whereby we fancifully see images of familiar things where they shouldn't exist – for instance, the face of Princess Diana in a slice of ham.

those of history's most bloodthirsty dictators, might one day wield extraordinary technological power.

Kokotajlo had the shortest timeline, as AGI predictions are called, of anyone at the think tank. 'In some ways,' he told me in 2024, 'being at CLR might have let me down easy with respect to the psychological change, because at CLR we thought about way worse things. Something like my timeline shortening wasn't at all the scariest thing I was thinking about.'

In 2021, Kokotajlo posted on LessWrong a set of predictions for the forthcoming years.[3] He correctly called the improvements to chatbots in 2022 and the big step-up in 2023 that hoovered up the chip output of NVIDIA but seemed not to translate into better self-driving cars. He expected an apparent slowdown in 2024, but an increase in the use of LLMs for propaganda via social media. He also predicted the sudden shortening of experts' AGI timelines. In 2025, according to this scenario, computing costs fall further; in 2026, AIs are better-equipped to carry out medium-term tasks, such as performing the role of an assistant or even friend. The stock market is 'going crazy', there's a new chip shortage and the question of AI sentience is becoming increasingly vexed. The predictions end there.

Kokotajlo was talent-scouted by OpenAI, who hired him in 2022. He joined Brundage's team as a governance researcher – someone whose job was to think about how the company and society ought best to manage AGI. Kokotajlo hoped to be part of a culture of responsibility and was encouraged by various gestures the company made towards safety. Privately, though, he remained worried about the speed at which OpenAI was rocketing towards AGI. He and his wife had had a daughter, but Kokotajlo felt that to have another child would not make sense. To have a child, he told me, is to indulge those dreams about teaching them this and that, sending them to college, welcoming them home for Christmas. 'And those dreams sort of crumbled, because they no longer felt so real to me.

'In the back of my mind would be: "Yeah, but that's not really gonna happen." Like, AGI is going to come probably before my daughter is a teenager, and then maybe it's going to go really well, and maybe it's going to go really poorly. But one way or another, she's not going to go to college,' he said, pausing to chuckle at the absurdity of the idea. Disaster, in his view, is more likely than not. 'The normie lifestyle, the normie plan … it's just probably not going to happen.'

AGI means different things to different people, and there are reasonable arguments to the effect that logistical bottlenecks* might slow things down a little. Likewise, it is much more challenging to create AIs that are useful in long-term, dynamic situations than it is to create AIs that spit out instant answers to single questions. Surveyed in 2022, a group of superforecasters thought AGI was 50 per cent likely to arrive by 2070.[4] (Superforecasters are analysts with a record of making accurate and quantified predictions. AGI is hard to quantify, so this was the yardstick they used: whether Nick Bostrom would call it AGI.[5])

Co-founders of each of the leading three AI labs, though, have predicted that AGI will arrive before the decade is out. In 2023, Altman teased Reddit users by claiming, and then denying, that 'AGI has been achieved internally'.[6] It might be in the interest of the labs to stoke hyperbole, but the view that AGI is imminent is not particularly controversial among AI specialists in the Bay Area of California. This is the social epicentre of the development of AGI, the place where the technology is being both accelerated and resisted. Here, it is taken as a given that AGI will happen fairly soon and that it will change our lives profoundly. Kokotajlo is simply behaving as if this is true.

He took his concerns to Sam Altman, OpenAI's CEO, telling him it was time for a 'pivot to safety'. By Kokotajlo's account, Altman agreed with everything he said. 'We'll do some cool stuff

* Energy, for instance.

soon,' said Altman, and, sure enough, OpenAI soon announced that a new 'Superalignment' team would be allocated 20 per cent of the organisation's computational resources. It was not as much as Kokotajlo wanted, but he was happy to see OpenAI putting more emphasis on safety.

'But then, over the course of 2023, it became clear that it was a bit of a gesture, and it wasn't anything close to what I had wanted. And then there was the board crisis and the aftermath in which all that progress got reversed.'

The board crisis took place in November 2023. After a year of astonishing progress, Altman was fired. The decision had been made by OpenAI's board, which had been put in place to ensure that the organisation stayed true to its mission of developing AI for the benefit of all humanity. Among the board's four members was Helen Toner, one of the co-authors of that paper published back in 2018. Altman had not been 'consistently candid', read a statement from Toner and her fellow board members. Toner, whose background is in effective altruism, would later say[7] that Altman lied repeatedly to the board, including about his ownership of OpenAI's start-up fund, that he gave inaccurate information about OpenAI's safety processes and that he was frequently manipulative, playing different parties against each other in order to buttress his own position.

This was the Silicon Valley equivalent of the assassination of a head of state. For a moment, it seemed that the effective altruists were going to be able to restore a culture of safety to OpenAI. But Altman, shoved off a precipice and swinging one-handed from the ledge, hauled himself back up. First, he was offered a senior role developing AGI for Microsoft, sparking predictions that Microsoft, with Altman directing a huge budget, would now overtake OpenAI. Then arose a groundswell of support for Altman from OpenAI staff and from investors. Within four days of being fired, Altman was reinstated. Toner resigned and the coup was over. Six months later, the Superalignment team was disbanded.

Kokotajlo had seen enough. Altman, he concluded, had been dishonest to him, too. 'This is how he gets loyalty ... he makes you feel like he agrees with you.' After much discussion with his wife, Kokotajlo decided to resign. OpenAI, in his view, was neglecting its sacred duty to develop AI safely, and this neglectfulness risked the future of humanity. (OpenAI did not respond to my requests for comment, but we can imagine that the Altman camp believes the stakes to be lower than did the ousted Effective Altruists.)

It turned out that resignation would be a little trickier than Kokotajlo had expected. He would be allowed to keep the OpenAI equity that was worth about $1.7m – the vast majority of Kokotajlo's net worth – but only if he agreed to sign a document requiring him not to be critical of the company. Kokotajlo and his wife had earmarked that money for charity and as savings, and their friends and family advised them not to jeopardise it. 'It was very tempting,' Kokotajlo told me, 'because people were correctly saying things like, "You can just sign it and then disparage the company anyway, and they're not going to sue you, because it would look so bad."'

But still he refused to sign the document. After a month of back-and-forth, he formally declined. He showed the email chain to a few allies and one of them leaked it to the online magazine *Vox*. Kokotajlo and the equity clawback were now the centre of attention. Overnight, the world learned what Kokotajlo had sacrificed in order to be frank about what he viewed as an intolerable risk of catastrophe. To those who were worried about the risks of imminent AGI, Kokotajlo was a paragon of quiet courage – and a cause for rare optimism. 'First time in your life you need to update your timelines,' went a tongue-in-cheek encomium, 'and the reason they changed is because the world sees you as a hero.'[8]

Altman himself responded to the equity clawback. The CEO said on X that he was 'embarrassed' by the potential equity cancellation, and that his staff were 'already in the process of fixing the standard exit paperwork'.

Kokotajlo's equity was restored, but it did not secure his silence. With five former OpenAI colleagues, and six who anonymously remained, Kokotajlo co-signed an open letter calling for frontier AI companies to commit to providing legal protection to whistleblowers. He also revealed a case of OpenAI breaking its own safety rules. Ahead of its official release, engineers had rolled out GPT-4 via an Indian iteration of Bing, the Microsoft chatbot. (OpenAI and Microsoft have a partnership.) This rollout broke an internal rule about getting approval for a safety board before conducting such tests. Referring to Kokotajlo's account, the *New York Times* ran the story. Microsoft first denied the test had happened, then admitted it had. Kokotajlo told the newspaper that OpenAI was 'recklessly racing' to be the first to AGI. And AGI, he estimated, was 70 per cent likely to be reached by 2027.

Kokotajlo and I were speaking towards the end of 2024. We were in a meeting room overlooking San Francisco, where the race for AGI was continuing unabated. OpenAI, founded as a non-profit, was transforming into a for-profit corporation.*[9] Attempts to formally regulate the frontier labs, with legislation rather than opt-in evals, were foundering. A few weeks prior, Kokotajlo and William Saunders, a fellow OpenAI whistleblower, had publicly called on Gavin Newsom, the governor of California, to sign into law the bill that would have imposed liability on the developers of dangerous models. 'Sam Altman, our former boss, has repeatedly called for AI regulation,' they wrote. 'Now, when actual regulation is on the table, he opposes it.' Kokotajlo and Saunders referred to OpenAI's demand for perpetual non-disparagement, on pain of loss of equity; the company's premature deployment of

* Its justification is that building AGI will take more resources than the company had previously imagined.[10]

GPT-4 in India; the threats made by the Bing chatbot to a reporter; the exodus of many other safety researchers, including Jan Leike, the FHI-er who helped devise RLHF, and the company's co-founder, Ilya Sutskever;* and a major security breach that the company did not disclose to the police or the government. 'While these incidents did not cause catastrophic harms,' wrote Kokotajlo and Saunders, 'that's only because truly dangerous systems have not yet been built, not because companies have safety processes that could handle truly dangerous systems.'

Kokotajlo's timelines are among the shortest in the game. Some experts, in contrast, still believe that the development of true AGI, let alone superintelligent AI, depends on breakthroughs not yet achieved. And Kokotajlo's prior accuracy, of course, is no guarantee of future accuracy. Nevertheless, some of Kokotajlo's fellow critics of frontier AI companies now see the situation as hopeless. Eliezer Yudkowsky, the Extropian and essayist who was one of the first to give serious intellectual consideration to the idea of AI alignment, believes doom to be near-certain. Informed by similar thinking, the campaigners of the Pause AI group are lobbying for a freeze on the development of powerful AI until we know how to build it safely and keep it under democratic control. Wearing their trademark orange T-shirts, Pause AI's volunteers stand outside frontier labs and the international summits that have followed the Bletchley gathering. But where those volunteers are calling for a pause, Yudkowsky wants to shut it all down, permanently. 'If intelligence says that a country outside the agreement is building a GPU cluster,' he wrote in *TIME*, 'be less scared of a shooting conflict between nations than of the moratorium being violated; be willing to destroy a rogue datacenter by airstrike.'[11]

* Earlier in the book, I described the succession of frontier labs founded on the premise that the existing labs were unworthy of being first to AGI. Sutskever's new venture, Safe Superintelligence Inc., is the latest.

Kokotajlo has the consolation of a life that is rich in meaning. 'I have hobbies, a beautiful wife, a lovely daughter. Life is pretty good, all things considered.' Nevertheless, he sees ours as a bleak era in human history; perhaps the final era. 'I've been dealing with this for several years,' he told me, 'so I'm sort of used to it. It definitely is quite depressing and sad.'

To live with these convictions, he said, is a little like what it must be to live in a country that is slowly but inexorably losing a war. You read the newspaper reports from the front lines; you begin to suspect that your side is, in fact, losing; you try to reason with the people around you who are caught up in the propaganda. It might be several years until you see the foreign troops marching down your streets, but every day they get closer.

With all this in mind, you might be wondering, having read my adumbrated version of Kokotajlo's predictions, what happens next. When Kokotajlo and I spoke, he and a small team of researchers were working on a book-length scenario that covers the next few years.

It is a scenario with two endings. One of them is a happy one. We figure out how to solve alignment, which means that the AIs act in our best interests and are honest with us. We introduce regulation that prevents companies from racing. There remain other problems, such as the race with China and the concentration of power in the hands of presidents or CEOs. But, overall, the AGI revolution is harnessed for the benefit of all humanity.

Yet there is another ending: the bad one. Here, we give the AIs increasing power. Just as primitive AI of the early 2010s helped social media companies produce more lucrative products, so the more sophisticated AIs of the late 2020s are embedded into corporations, militaries, public services and the development of new AIs. But in this scenario, the work I've described in these chapters has been unsuccessful. Mechanistic interpretability didn't work, and our control efforts didn't work, and evals organisations didn't spot a new danger. The whistle was blown, but nobody heard. 'Whatever alignment techniques we had used,' explained Kokotajlo, 'just failed. And we didn't notice it, because we weren't paying enough attention, because they' – the AIs – 'are black boxes.

'Oops,' he said. 'Turns out they weren't aligned. Now we're fucked.'

CHAPTER 14

THE METHUSELARITY

'Since then – 'tis Centuries – and yet
Feels shorter than the Day
I first surmised the Horses' Heads
Were toward Eternity –'

Emily Dickinson, 'Because I could not stop for Death'

Let's assume that everything goes well, that our species makes it to the year 2100. No city-sized asteroids have come hurtling out of the darkness of space. Although our skies have been illuminated by a few solar storms, our power grids withstood them. From the mid-2020s, we began to cooperate on solar geoengineering, even as we improved our collaboration on pandemic prevention. As a result, the 21st century turned out to be a bad one for pathogens, which we found ever-more inventive ways of massacring, and a good one, as it eventually turned out, for polar bears. Perhaps a modest volcanic eruption stimulated us into cooperating, pan-nationally, on a plan to investigate the defusing of supervolcanoes.

An advanced AI had crashed the stock market in 2029, resulting in financial chaos in the short term and well-calibrated regulation in the medium term. Governments inspected one another's biggest data centres as they did one another's nuclear weapons silos, leveraging a series of technical breakthroughs to ensure that AIs were aligned, controlled and hailed as responsible for an era of unrecognisable prosperity. As governments deliberated on which great problem to vanquish next, it was noted with some relief that no human had yet been turned into a paperclip. (Meanwhile, to the delight of geothermal fans, the Kola Superdeep Borehole had long since been surpassed.)

It is difficult, of course, to describe a world transformed by artificial general intelligence, let alone one that has uncorked artificial super-intelligence. Such a world might be unrecognisable. Even in that world, though, things would have to go extremely well for you and I to stand a good chance of making it to the year 2100. If you are currently an adult, today's projections of life expectancy suggest that the odds are against you. If you happen to be a small child,* then you might have a decent shot. But by 2100, you will be withered and fading. 'In the long run,' as John Maynard Keynes put it, 'we are all dead.'[1]

Conventional taxonomies of catastrophic risks tend not to include ageing. Perhaps they should. The vast majority of human deaths result from diseases to which we are all but immune in our youth; but with age comes vulnerability. Ageing, then, is the peril that will come for all of us – even if we see off every other menace. As Nick Bostrom observed in 2008: 'If ageing were not certain but merely probable, it would immediately shoot to the top of any list of global catastrophic risks.'[2]

And there is increasingly solid justification for believing that ageing might end up being just that: probable rather than certain. Yet that certainty is deeply psychologically ingrained in us. So accustomed are we to death's inevitability that you might already be bridling at the thought of a life extended by science. You don't *want* to live forever, you wouldn't want to be a burden, death is only natural, the world is crowded enough as it is. These are some of the most commonly voiced objections to

* Perhaps you have strayed deep into this book in the hope that it will be about superheroes, rather than technical researchers. If so, I can only apologise.

the science of longevity. We vigorously resist almost any other threat to human life, yet ageing is regarded as somehow sacred. In one of his fables, Bostrom compares it to a dragon that tyrannises a whole world. It is a dragon whose defeat is so inconceivable that spiritual leaders presented being eaten by it as good and natural. 'They said that it was part of the very meaning of being human,' Bostrom writes, 'to end up in the dragon's stomach. Others still maintained that the dragon was good for the human species because it kept the population size down.'

Not everyone is of that kind of mind. To certain kinds of scientist, extended life is a beguiling prospect: part engineering problem, part grail quest. But their predecessors in this quest were quacks. Ageing is a deeply complex phenomenon, enacted from individual cells via countless different biological processes. It is not to be defeated by bathing in the blood of the young, finding just the right combination of herbs, or grafting onto oneself the testicles of a monkey.[3]

If we are to defeat ageing, we must understand it rather than attempt to second-guess it. And by the end of the 20th century, we were making some progress in this respect. It was halting, expensive progress, but it was progress all the same. An important catalyst was the Human Genome Project. Over a period of 23 years, the project's scientists worked out how three billion pairs of chemicals form the 20,000–25,000 genes that define a human. This, roughly, is what is meant by DNA sequencing.

The most immediate result of the project was that scientists acquired their deepest understanding yet of human genetics. But there was another, more diffuse result: a huge reduction in the cost of DNA sequencing. That reduction in cost has made us more vulnerable to man-made pandemics, but it has also encouraged scientific inquiry into the most fundamental workings of the human body. We know which genes predispose us towards Alzheimer's, for instance. Consumers can send off a cheek swab and receive an analysis of their genome in return.

And with the help of a new technology, CRISPR, scientists are able not only to sequence a genome, but to insert new genes. First, we read the book of life; now, we can edit it.

Before these advances in genetics, there were few scientists willing to take on the dragon. One of that small number was Aubrey de Grey, an Old Harrovian computer scientist who became a gerontology researcher and anti-ageing crusader.* De Grey's prophecies – notably, 'the first human to live to 1,000 has already been born' – are lent additional weight by their being uttered through a long beard. (Depending on their level of sympathy for de Grey, writers compare this beard to that of an Old Testament prophet or of Rasputin.[4]) In 2003, de Grey co-founded a research institute, the Methuselah Foundation, whose goal was to create life extension technologies. Some futurists envisage the 'Singularity', where super-intelligent AI allows man to elide with machine; de Grey looks forward to the 'Methuselarity', where new anti-ageing therapies will enable us to live long enough to benefit from the next generation of such therapies. This is the point at which we will have reached, as de Grey puts it, 'longevity escape velocity'.

De Grey conceptualises ageing as the cumulative effect of damage to cells and has offered prizes to research teams that extend the lifespan of mice.[5] It is mice that are most likely to be the subjects of longevity research – mice, nematode worms and fruit flies. (Mice are mammalian and easy to maintain; nematode worms and fruit flies have simple brains, making them good starting points for study.)

The advances in genetics encouraged more scientists to join the field, with investors offering support. The amount spent on longevity is minuscule compared to the amount spent on its effects, such as cancer and dementia. But it is enough to have

* It is rare, in science, to move from one disciple to another. Members of the anti-catastrophe league are much more likely than their peers to have undergone these intellectual transformations.

funded several new institutes, including the Google spin-out Calico and the Sam Altman-backed Retro Biosciences. In all, longevity has become a field not unlike nuclear physics in the 1930s. For inter-war Cambridge, read 2020s Silicon Valley, where a small group of ambitious researchers are trying to work out how to turn a young and speculative science into a tremendous force for good.

Martin Borch Jensen is one of those researchers. As a boy, Jensen was curious about life's meaning. Among the writers and books he turned to were the French existentialists and *Bhagavad Gita*. The latter is a scripture that describes a prince, Arjuna, being encouraged to release himself from desire and to accept his place in the cosmic order. In Jensen's view, the *Gita* was 'exactly wrong'. The young Dane felt that the urge for progress, the act of wanting, was 'the essence of life'. No philosopher seemed to Jensen to have identified exactly and completely what is meaningful in life, but life itself seemed a good start. As Jensen told me, years later, life 'is at least, if not sufficient, then at least necessary for meaning in life'.

He took a PhD in the role of DNA repair in the ageing process, subsequently moving to California to join a longevity research lab, the Buck Institute. On finishing his postdoc, Jensen was offered a sizable grant by the National Institutes of Health, which allocates US government money to medical research. But he turned it down. 'If I spent the next seven years doing what I just got a grant to do, I wouldn't have the most possible impact on ageing.'

He decided that founding a start-up would allow him to move faster. Borch was introduced to Francisco LePort, a former Tesla research engineer with a background in physics. LePort had been a child prodigy, starting university at the age of 14. Soon he was helping his mother, Cristina Rizza, with academic work on the longevity of – naturally – fruit flies. Rizza was a cardiologist, novelist and biotechnology entrepreneur, and she would later help create a strain of healthy, vigorous 'Methuselah' fruit flies that live three times as long as their peers.[6] All this had been enabled by advances in genetic and genomic sequencing. LePort was busy founding a software company, but his interest in his mother's work was sharpened by the loss of his grandparents. Nor could he ignore the ageing of his parents. As he would later put it, LePort was 'becoming more and more aware of where it was all going for myself, my kids, and anyone I cared about'.[7]

Jensen and LePort met at a bakery. It went well enough to merit another meeting. Here, Jensen questioned LePort on his departure from his software company: 'Are you a quitter?' LePort answered the question seriously and without defensiveness. It turned out to be the right approach; Jensen was impressed. The courtship continued with a trip to Burning Man, a techno-utopian festival on the harsh alkaline flats of Nevada's Black Rock Desert. 'We'd only really been discussing things logically,' recounted Jensen, 'which doesn't say much about how we'd each react when things got messy.'[8]

Jensen and LePort knew they wanted to work together on longevity, but they still hadn't decided how they would do it. In the end, the idea came from LePort's mother's fruit flies. Thanks to further advances in genetics, it was now possible to take individual genes from a healthy animal (via a procedure called single-cell sequencing) and to insert them into the diseased organs of another fly (a procedure known as gene therapy). Jensen and LePort agreed to combine these two advances, applying them to animals such as horses, which are genetically closer to humans than are flies.

The logic was this: to defeat the diseases that relate to age, they would use the genes via which younger animals stay healthy. And they would spread their bets. The odds of a given gene therapy being successful were low, but Gordian Biotechnology – Jensen and LePort named their new company after the maddeningly complex knot of Ancient Greek legend – would test hundreds of therapies per sick animal. Gene therapy would allow them to precisely edit the genetic code of living cells; single-cell sequencing would enable them to closely monitor the small-scale effects of that editing; and machine learning, the method that enabled the advances in AI that I discussed in the previous chapter, would help the researchers understand how changes those small-scale effects would affect a living animal. Jensen and LePort hoped that, in this way, they could prepare new treatments for human trials more quickly than had ever been previously possible.

In October 2024, I visited the Gordian headquarters, an office and set of laboratories in San Francisco. (There were no stables, disappointingly – the horses are kept in Florida.) Outside the building was a Cybertruck, while inside were white-coated technicians, sci-fi posters and so many big white blocks of fridges and lab machinery that the place felt like a very high-tech laundromat.

It was almost exactly six years after the company was founded. Defeating age-related disease is slow, expensive work, but Jensen and LePort had made progress. Their earliest work had targeted metabolic dysfunction-associated steatohepatitis, better known as MASH, or non-alcoholic liver disease. The older and fatter you get, the more likely you are to silently acquire MASH. And the more advanced your case of MASH, the more likely you are to develop fatigue, liver cirrhosis and several forms of cancer. Three million Americans are afflicted with this disease.

Jensen and LePort aimed to prove that their methodology could predict the outcome of several gene therapy trials at once, in a single animal. As such, they introduced 25 gene therapies into a mouse suffering from MASH, evaluating the result with their array of organ-monitoring technology. In early 2021, the results came through. Gordian had successfully predicted 13 out of 16 clinical outcomes. This success earned them enough investment to triple the size of the team, demonstrate proof-of-concept for their precise monitoring of animal organs and lay out strategies for ailments such as heart failure and osteoarthritis. At the time of my visit, they were receiving the first sets of data from the treatments they have been giving their animals – data that should guide a raft of clinical trials. The best-case outcome will be that Jensen and LePort will help shepherd onto shelves an array of treatments for the diseases through which age kills us. Jensen believes they are working on 'the biggest lever for human health in this century', one that makes other work seem trivial. When talking about ageing, be it his own, his parents' or that of other people, he has a stark lucidity to him. I believed him when he told me the following: 'It is challenging for me to find it very meaningful to do something else knowing that my entire existence will vanish.'

In a world where ageing is the dragon of Bostrom fable, rather than bodily degeneration, Jensen's work would be the defanging of this beast, the sanding-down of its claws, the blunting of its teeth, the surgical removal of whatever organ in the dragon's throat produces jets of fire. Any diminution of the dragon's powers would be a tremendous result.

But there are some scientists whose goal is not just to denude the dragon of some of its most fearsome weapons, but to destroy it entirely – to end ageing. Chris Bradley is one of those scientists and his work in this field has been enabled by the sale, in 2018, of the healthcare data company he founded in 2012. 'I can do what I always wanted,' he realised, 'which is to chase the hardest, most complex problem I can think of that I still have a chance to do something about.'

The problem was one that he had been thinking about since, as a student, he read de Grey's writing on longevity escape velocity. It was that of ageing, which, for the first time, was no longer an alchemical problem but a scientific one. Rather than decamp to a beach, Bradley founded a new company – one that would, like Gordian, make use of new technology to solve an ancient problem.

At this point in scientific history, though, we understand ageing so poorly that there is not yet consensus on how to measure it, let alone what causes it. Some scientists have pointed to telomeres, the structures that protect a cell's genetic material. When they shorten, we age. Or is it the other way round? We still don't know.

Bradley's bet is that ageing comes from the DNA itself. It is energy-intensive to repair DNA, which is perhaps why our bodies get less good at it as we age. Instead, we expend energy on survival. (In contrast, animals that have spent hundreds of thousands of years experiencing less intense selection pressure – sharks, whales, naked mole rats – can live for much longer than nature usually permits.) But errors in DNA transmission cause the little mutations in cells that, as the years take their course, snowball into larger and more tangible effects: grey hair, malfunctioning hearts, forgetful minds. 'Your car slowly gets damaged in different places,' Bradley told me, 'and eventually breaks down.'

At Matter Bio, which is the name of Bradley's New York-based company, the approach is to detect the loss of genomic information and correct those mutations via editing the genome of living beings.

Some cells, though, are too decrepit to keep. Bradley aims to replace them 'not with their own copy, but with a better version. I'm going to replace it anyway. Might as well give you an updated car, right? You get the Corvette upgrade.'

It is one thing to insert a gene into a virus. It is another to insert it into a living being. Biotechnology companies tend to be cagey about their methods, but one method uses viruses themselves – engineered viruses – as a way of transferring the material into a live body. Another uses intravenous injection to insert into the bloodstream some fatty particles that eventually reach the organ of choice. Neither of these methods is yet perfected. Delivery is one of many challenges faced by Bradley and his peers in the field. If he succeeds, though, he might be able, he believes, to add 30 years of health to a human lifespan.

From there, our best bet might be whole-brain emulation. This is the proposed technology that Sandberg and Bostrom puzzled over in the early days of the Future of Humanity Institute. It is the method by which we might make a digital copy of our physical brains, freeing us from the limitations of our mortal bodies. If humans are to live for aeons, see new galaxies and access realms of experience far beyond our current imagination, it will be via whole-brain emulation. This technology is a long way from fruition; the physical structures that form our memories and personalities are impossibly small and therefore near-impossible to scan and simulate. Some progress has been made, though. Weeks before I spoke to Bradley, a group of 36 scientists announced that they had successfully simulated on a laptop the brain of a fruit fly.

Whole-brain emulation presents difficult questions about identity. Is a scanned version of your brain the same being as you? What sort of digital lives would we want to live? Would we want to live in a new digital world, or would we want to be downloaded straight back into a new body? Given the difficulty of developing this technology, we will probably have many years to consider these questions.

In the meantime, scientists such as Jensen and Bradley might eke us a few more years, just as modern medicine has added a few per cent to adult life expectancies. Perhaps any amount of additional longevity sounds dystopian to you. But death, too, is dystopian. You, I and everyone we love is destined for oblivion, and we are likely to be dragged there in pain and against our will. If your concern is that rich people might be first to access longevity therapeutics, then why must the answer be death for all? Perhaps you feel there are too many people on this planet. But why must the answer be doom? The answer to pollution is the remedying of it, not the infliction of oblivion on everyone in the world. And if death gives life meaning, then why don't we want to live shorter lives? Why should the anti-catastrophe league turn up to work?

You will recall the name of Eliezer Yudkowsky, the Extropian whose writings helped seed the development of AI safety. In 2004, Yudkowsky lost his younger brother, Yehuda, to a devastatingly premature death. Yehuda was 19. As Yudkowsky put it in an email to some fellow transhumanists, Yehuda 'is not "resting in peace". Yehuda is not coming back. Yehuda doesn't exist any more. Yehuda was absolutely annihilated at the age of nineteen.'

The poet Dylan Thomas urged his father to 'rage against the dying of the light'. Yudkowsky raged against its extinguishing. 'Nineteen years, 7,053 days, of life and memory annihilated. A thousand years, or a million millennia, or a forever, of future life lost. The sun should have dimmed when Yehuda died, and a chill wind blown in every place that sentient beings gather, to tell us that our number was diminished by one.'

CONCLUSION
EXISTENTIAL HOPE

Power's script, – wound, bobbin-bound, refined –
Is stropped to the slap of belts on booming spools, spurred
Into the bulging bouillon, harnessed jelly of the stars.
Towards what?

Hart Crane

Those in favour of vastly extended human lifespans must reckon with one uncomfortable truth: some deaths are worth wishing for. Think of the tyrants of the 20th century and beyond. Once Stalin was dead, many fewer Russians were purged each year. Similarly, the most murderous excesses of the Chinese Communist Party ended with Mao. These regimes, and regimes like them, are responsible for more human suffering than almost any of the phenomena I have described in this book.

It is troubling, therefore, to imagine a world in which Kim Jong-Un, for instance, might live many times longer than he can currently expect to. In such an eventuality, the people of North Korea would have to endure many more centuries of despotism. And there have been more than enough despots for us to know that tyranny might be part of humanity's future as well as its present and past. If that future contains long-lived tyrants who have sole control over powerful artificial intelligence, then it is hardly a future to look forward to.

This book has been about existential risks, sometimes known as 'X-risks'. The scenario of the long-lived tyrant constitutes what certain philosophers call an 'S-risk': a risk of extreme suffering on an extreme scale. Locked-in tyranny, as this eventuality is called, is not, in itself, an argument against longevity, AI and progress. Some observers, such as the DNA repairman Chris Bradley, are optimistic about humanity's prospects. In his view, we might manage ourselves via digitised forms of governance that, much like cryptocurrency networks, cannot easily be altered once set in motion.

So perhaps there are ways of averting tyranny, even though it seems a recurrent feature of human society and even though its risk could be exacerbated by technology, rather than lessened by it. Nevertheless, we must not ignore the possibility of there existing, now or in the future, suffering so extreme that it might render worthless the labours of all those I have written about in this book. Some S-risks are more insidious than we might imagine; consider the hundred billion animals that are currently enduring the unimaginable misery of factory farming. As far as those animals are concerned, an unnatural hell already exists.

And perhaps we might create a similar fate for AIs. Our empathetic faculties were designed to be receptive to our fellow humans, rather than to beings that do not have in common with us even the fact of being carbon-based. Perhaps reinforcement learning, the kind of training that involves a rap on the digital knuckles, might inflict grievous suffering on AIs of the future, the kind of AIs that will be more powerful and sensitive than those we currently deploy. And to such an AI, it might seem abominable to be conjured into existence only briefly, before being annihilated once we have finished asking it the capital of Bolivia. Such is the fate of each instantiation of ChatGPT. It is unlikely that ChatGPT can feel anything, let alone existential angst, but its successors might.

S-risks, therefore, must give pause to any scholar of X-risks. Some deaths are worth wishing for, and some futures are worth fearing. There is little point labouring to avert a sudden catastrophe if the consequence is a slow-burn moral disaster.

What, then, is the point of preserving our world? What is the point of the anti-catastrophe league? Almost everyone, perhaps bar some particularly hardcore environmentalists and cultists, would agree that it is better that we exist than do not. We can all be silently thankful that Stanislav Petrov did not initiate a nuclear holocaust. But perhaps we can do better than perpetuate a status quo that, in many respects, is in such urgent need of improvement. If so, perhaps there is reason to feel excitement about the future as well as trepidation.

The Future of Humanity Institute brought attention to many perils. As it helped demonstrate, humanity has opened a Pandora's box. Curiosity beguiled us into lifting its lid, and all manner of monsters rushed out. They cannot be put back in. The Greek myth has it that before the box was slammed shut, out fluttered hope. And there is a version of this in apocalypse lore: a concept called 'existential hope'.

In 2015, Owen Cotton-Barratt and Toby Ord wrote a brief report that introduced the idea.[1] In doing so, they drew on a word coined by J. R. R. Tolkien, the author of *The Lord of the Rings*. The word was 'eucatastrophe', a fusion of one ancient Greek word ('catastrophe', in its sense of a sudden change of circumstances) with another (the prefix 'eu', meaning 'good').

'It is the mark of a good fairy-story, of the higher or more complete kind,' Tolkien wrote, 'that however wild its events, however fantastic or terrible the adventures, it can give to child or man that hears it, when the "turn" comes, a catch of the breath, a beat and lifting of the heart, near to (or indeed accompanied by) tears, as keen as that given by any form of literary art, and having a peculiar quality.

'Even modern fairy-stories,' Tolkien concluded generously, 'can produce this effect sometimes.'[2] So can the real world. A 'turn' of the kind to which Tolkien refers is discernible in the Enlightenment, the victory by the Allies in the Second World

War, and perhaps even the replacement, via the arrival of Chicxulub asteroid, of dinosaurs with mammals.[3]

And so Cotton-Barratt and Ord, inspired by Tolkien, came up with an antonym for the baleful term that is 'existential risk'. They wanted 'to be able to refer to the chance of an existential eucatastrophe; upside risk on a large scale. We could call such a chance an existential hope.'

They continued: 'Some people are trying to identify and avert specific threats to our future – reducing existential risk. Others are trying to steer us towards a world where we are robustly well-prepared to face whatever obstacles come – they are seeking to increase existential hope.'

The first group, those trying to identify and avert those specific threats, have been the principal focus of this book. The second group, those trying to increase our quotient of existential hope, often overlaps with the first. The same FHI staff who wrote about the most disastrous scenarios for life on Earth would also envisage the opposite. Nick Bostrom's latest book depicts a world in which humans find meaning in a world where super-intelligence has solved almost every problem. Anders Sandberg's explores the maximal potential of a space-faring civilisation.

'One amazing future,' said Sandberg in 2023,[4] 'is that humanity gets its act together. It solves existential risk, develops molecular nanotechnology and atomically precise manufacturing, masters biotechnology, and turns itself sustainable: [it] turns half of the planet into a wilderness preserve that can evolve on its own, keeping to the other half where you have high material standards in a totally sustainable way that can keep on going essentially as long as the biosphere is going.'

Beyond the biosphere, there might be ways that we can sustain ourselves for many more billions of years from the present day. 'The best method, in my opinion,' said Sandberg, 'is to use black holes. I'm very fond of black hole power. And I am assuming that maybe, in a few trillion years, I'm going to be dealing with protesters saying, "No black holes in our neighbourhood" and "Don't build that power plant, Anders". But they're actually lovely.'

Throw a black hole some matter, such as spent stars, and it will release radiation that you can put to good use. It is a vast, cosmic recycling plant. Bostrom, you might recall, wrote about 'astronomical waste', the gargantuan amounts of energy that, every second of every day, is being flushed down the great cosmic lavatory. At some point in the future, we might harness this energy from black holes and elsewhere – or at least however much of that energy is still in reach as the universe expands.

It is difficult to imagine our being able to reverse that expansion, or that we will be able to prevent the end of the stelliferous period: the current, multi-billion-year era in which the universe is dotted with, in the form of stars, free fusion reactors. It is harder still to imagine that we might be able to overcome entropy, the universe's natural tendency towards degradation. As we move further in time from the Big Bang, stars and galaxies become ever-more sundered, with ever-fewer stars being formed. Eventually, matter will be so thinly spread that the universe will reach a state of heat death, an eternity of cold and darkness. Perhaps the greatest conceivable challenge for our descendants is to avert that fate.

These futuristic visions, though appealing, are too far away and too abstract to motivate most people today. But there are more modest forms of existential hope to be found everywhere – even alongside the risks I have written about in this book. The enriched uranium rescued by Andy Weber, for instance, was repurposed as fuel for nuclear power stations. And newer technologies will bestow even broader benefits. Just as our new aptitude for gene editing might one day give us more years of health and happiness, so our development of artificial intelligence might enable us to make our lot much better, even within our lifetimes. With the help of problem-solving intelligence, we could cure cancer, end mental illness, and cut away other blights on our happiness; we could devise new technological methods by which to reverse the damage we have done to our planet; we could equip ourselves to understand the secrets of the universe and have experiences beyond our current imagining. To protect ourselves from existential risks is to enable feats of civilisational greatness.*

And should we ever build for ourselves a future worthy of existential hope, a future full of great feats and meaningful lives, we will have much to owe to the anti-catastrophe league. In an important respect, they are swimming upstream. They are resisting the impersonal forces that are shaping our world. As I have argued, proliferation of modern risk seems, in many ways, to be systemic; those who expose us to catastrophe are often acting in keeping with powerful fields of incentives. Develop super-intelligent AI, for instance, and you will become wildly powerful. Create a nuclear super-weapon and you might ensure the safety of your country, even if the world is less safe as a result.

To reduce those risks, however, requires defiance of those structures of incentives. Thick-skinned iconoclasts, such as the early asteroid truthers, draw attention to a threat. Brave individuals, such as Stanislav Petrov, defy wrong-headed orders. Swashbuckling diplomats, such as Andy Weber, seize the initiative in chaotic situations. Courageous whistleblowers, such as Daniel Kokotajlo, risk their family's financial security in order to advocate for the safe development of dangerous technology.

Others are attempting to rewrite those incentives altogether. Ingenious policymakers use carrots and sticks to change behaviour for the better, complementing the brilliant inventors who are rearranging the physical world for humanity's benefit.[5] The writer Rutger Bregman, meanwhile, calls for us to collectively assign higher social status to those who devote their careers to improving the safety or wellbeing of humanity. And my friends at the Meaning Alignment Institute are working on an ambitious

* These feats need not feel impersonal. I explore a British-flavoured futurism in *Anglofuturism*, the podcast I present with Calum Drysdale.

attempt to use AI to elicit from real humans the implicit web of values that underlies our behaviour. In partnership with OpenAI, they hope to instil this web of values into powerful AI systems, creating what they call 'wise AI'. This is the kind of AI that is nurturing rather than exploitative, insightful rather than insensitive, enriching rather than enfeebling. The institute aims not merely for AI that is aligned with human values, but at a society and world that is aligned in the same way. This would constitute what the institute's researchers have termed 'full-stack alignment'. Such a world is a distant dream, but it would entail the weakening of the perverse incentives that have stripped the modern world of meaningfulness even as it becomes more dangerous.

The wilful individuals I have described in this book are exceptional in more ways than one. They are admirable, but they are also unusual. Most people cooperate with the forces of social and economic gravity that have pulled us towards catastrophe. Only a few have the lucidity to understand what is at stake. And only a subset of that small group has shown the moral courage required to defy those gravitational forces. Many of them seem eccentric, but, in this regard, they are the sanest people among us. Their minds have caught up with the reality that our world is much grander, and more pliable, than the world for which evolution designed our brains.

I cannot deny that, in writing this book, I have been frequently dismayed by the crassness with which our species has repeatedly endangered itself. Nor am I confident that we have successfully enumerated every threat to us; more likely is that the universe continues to harbour hazards yet unseen.* But I have also been

* The universe remains mysterious. It is worth dwelling on Michael Collins' description of the experience of flying, alone, around the dark side of the Moon. Collins was the pilot who had dropped off Neil Amstrong and Buzz Aldrin for the first lunar landing. 'I am alone now, truly alone, and absolutely isolated from any known life. I am it. If a count were taken, the score would be three billion plus two over on the other side of the moon, and one plus God knows what on this side.'[6]

uplifted, now and then, by the efforts of those who are attempting to unpick that danger.

With that in mind, consider the theory that the universe appears devoid of life only because life always manages to eradicate itself. Given that context, one might expect humanity to become just another of those self-extinguishing species. Perhaps Earth will one day bear as little trace of our brief era as the Amazon does of the various tribes who came, went and left their temples and houses to be subsumed by the voracious rainforest. We know of these people's existence only via machine-learning analysis of images taken from satellites. Centuries of struggle, joy, peace and war – mighty deeds, great romances, brave feats – and now oblivion.

It is usually foolish to imagine oneself, or one's species, to be immune to the forces that propel everyone else to their destiny. Perhaps the same applies to humanity and existential risk. Are we somehow different, somehow immune? Will we become one of those long-forgotten Amazonian tribes? Will there be anyone left to behold whatever ruins we leave behind? Or will we flourish?

I don't know. Yet I come away from this work with the feeling that our destiny, for better or worse, seems rather malleable. It is like glowing metal that has been laid across a blacksmith's anvil. Hammers raised, we stand as smiths before that glowing steel. The shape it will take is ours to choose.

ACKNOWLEDGEMENTS

I am principally grateful to the hundreds of experts who have gamely explained their work to me. This book would not exist without their generosity. Many of those experts are quoted in these pages. Some are not. I thank them all.

Parts of this book have their origin in reporting I produced for five magazines: *Asimov*, *Asterisk*, *BBC Future*, *Prospect* and *Works in Progress*. I am grateful to my editors at each of these publications: Xander Balwit, Nico McCarty, Clara Collier, Jake Eaton, Richard Fisher, Richard Gray, Emily Lawford and Ben Southwood. I thank them not only for letting me repurpose my work, but for significantly improving it.

In a similar vein, this book draws on research I conducted for two think tanks: Blueprint Biosecurity and the Institute for Progress. I am grateful to David Carel, Santi Ruiz and Arnab Datta. They commissioned the two reports and they sharpened my thinking on pandemic prevention and clean energy.

I am also grateful to my former colleagues at Effective Giving. This was the philanthropic advice organisation whose work centred on many of the topics that this book deals with. I owe particular thanks to Megan Brown and Dilan Saujani.

I am indebted to the staff of Constellation, the collaborative research centre in Berkeley, California. The six months I spent there as a visiting fellow were immensely valuable to my production of this book.

Many other people have lent their time and insight to this project. I owe especial thanks to Joel Diggory, Ed Higson, Jason Hausenloy, Paul Ingram, Ben James, Joe Joseph, Lauren Mangla, Paul Murray, Max Nadeau, Latifah Salaudeen, Maeve Scullion, Zach Stein-Perlman, James Wagstaff, Jack Weston and Keith Wynroe. I am particularly grateful in this respect to Matt Kilcoyne. I ask the forgiveness of anyone I have forgotten.

For their able stewarding of this project, I am also grateful to my agent, Katie Fulford; to my editor at Mudlark, Joel Simons; and to their colleagues. In more diffuse ways, my work on this book has been supported by friends and family on either side of the Atlantic. My final thanks are to them.

NOTES

INTRODUCTION: DEFUSING DOOMSDAY

1. Ruiz, Santi, 'How to Secure Weapons-Grade Uranium', *Statecraft*, 11 October 2023. (https://www.statecraft.pub/p/how-to-secure-weapons-grade-uranium-d28)
2. Hoffman, David E. *The Dead Hand: Reagan, Gorbachev and the Untold Story of the Cold War Arms Race*, New York, Doubleday, 2009.

CHAPTER 1: THE DOUBLE ASTEROID REDIRECTION TEST

1. Marvin, U. B. 'Meteorites in History: An Overview from the Renaissance to the 20th Century', in McCall, G. J. H., Bowden A. J. & Howarth, R. J. (eds). *The History of Meteoritics and Key Meteorite Collections: Fireballs, Falls and Finds*, London, Geographical Society of London, 2006.
2. Lindsay, E. M. 'Maskelyne and Meteors', *Irish Astronomical Journal*, vol. 8(3) (1967), pp. 69–72.
3. Ibid.
4. Marvin, U. B. 'Ernst Florens Friedrich Chladni (1756–1827) and the Origins of Modern Meteorite Research', *Meteoritics & Planetary Science*, vol. 31 (1996), issue 5, pp. 545–588.
5. Ibid.
6. Marvin, 2006.
7. Biot, J-B. *The Meteorite Report: Relation of a Voyage Made to the Departement of Orne in Order to Verify the Reality of a Meteorite Observed on the 6th of Floreal of the Year 11 (April 26th, 1803)*, Princeton, Metron Publications, 2013.
8. Golia, M. 'Fire in the Sky: A History of Meteoritics', *Engelsberg Ideas*, 2021. (https://engelsbergideas.com/essays/fire-in-the-sky-a-history-of-meteoritics/)

9. Moynihan, T. *X-Risk: How Humanity Discovered its Own Extinction*, Urbanomic, 2020, p. 8.
10. Jha, A. 'It's Called Apophis. It's 390m Wide. And it Could Hit Earth in 31 Years' Time', *Guardian*, 7 December 2005. (https://www.theguardian.com/science/2005/dec/07/spaceexploration.research)
11. Lea, R, 'Apophis: The Infamous Asteroid We Thought Might Hit Us', Space.com, 14 December 2022. (https://www.space.com/apophis)
12. Senel, C. B., Kaskes, P., Temel, O. et al. 'Chicxulub Impact Winter Sustained by Fine Silicate Dust', *Nature Geoscience*, 16 (2023), pp. 1033–1040. (https://doi.org/10.1038/s41561-023-01290-4)
13. 'Near-Earth Object Survey and Deflection Analysis of Alternatives', NASA, March 2007. (https://www.nasa.gov/wp-content/uploads/2015/01/171331main_neo_report_march07.pdf?emrc=cb50d9)
14. Talbert, T. 'Five Years After the Chelyabinsk Meteor: NASA Leads Efforts in Planetary Defense', NASA, 15 February 2018. (https://www.nasa.gov/solar-system/five-years-after-the-chelyabinsk-meteor-nasa-leads-efforts-in-planetary-defense/)
15. NASA [@NASA], 'IMPACT SUCCESS! Watch from #DARTMission's DRACO Camera, as the Vending Machine-Sized Spacecraft Successfully Collides with Asteroid Dimorphos, Which Is the Size of a Football Stadium and Poses No Threat to Earth. https://T.Co/7bXipPkjWD', Twitter/X post, 27 September 2022. (https://x.com/NASA/status/1574539270987173903)
16. Dobrijevic, D. 'NASA's DART Asteroid-Smashing Mission: The Ultimate Guide', Space.com, 14 October 2022. (https://www.space.com/dart-asteroid-mission)
17. 'Double Asteroid Redirection Test (DART)', NASA, accessed 15 February 2025. (https://science.nasa.gov/planetary-defense-dart/)
18. 'Hubble Captures Movie of DART Asteroid Impact Debris', NASA, 1 March 2023. (https://science.nasa.gov/missions/hubble/hubble-captures-movie-of-dart-asteroid-impact-debris/)
19. Bardan, R. 'NASA Confirms DART Mission Impact Changed Asteroid's Motion in Space', NASA, 11 October 2022. (https://www.nasa.gov/news-release/nasa-confirms-dart-mission-impact-changed-asteroids-motion-in-space/)
20. 'Eros', NASA, accessed 15 February 2025. (https://science.nasa.gov/solar-system/asteroids/433-eros/)
21. 'Falling to Earth: The Chelyabinsk Meteorite', National Museums Scotland, accessed 15 February 2025. (https://www.nms.ac.uk/discover-catalogue/falling-to-earth-the-chelyabinsk-meteorite)

22. Lea, R. 'This is What Would Happen if Scientists Found an Asteroid Heading to Earth', Space.com, 6 April 2023. (https://www.space.com/nasa-models-hypothetical-asteroid-impact-scenario)
23. Sagan, C. and Ostro, S. J. 'Dangers of Asteroid Deflection', *Nature*, 368 (April 1994), p. 501. (https://doi.org/10.1038/368501a0)
24. Gibbings, A. 'Laser Bees: A Concept for Asteroid Deflection & Hazard Mitigation', University of Strathclyde Advanced Space Concepts Laboratory, accessed 15 February, 2025. (https://pure.strath.ac.uk/ws/portalfiles/portal/23588995/Gibbings_A_et_al_Pure_Laser_bees_A_concept_for_asteroid_deflection_and_hazard_mitigation_2012.pdf)

CHAPTER 2: A RAGING STAR

1. 'Aurora Borealis: A Brief Overview', US National Park Service, 26 October 2021. (https://www.nps.gov/articles/aurora.htm)
2. NASA Science Editorial Team, 'What is a Solar Flare?', NASA, 9 August 2011. (https://science.nasa.gov/science-research/heliophysics/space-weather/solar-flares/what-is-a-solar-flare/)
3. 'From the Northern Lights to Communication Breakdown: How Does Space Weather Affect the Earth?', Science Museum, 26 September 2018. (https://www.sciencemuseum.org.uk/objects-and-stories/how-does-space-weather-affect-earth)
4. 'Solar Storm Carried the Energy of 10 Billion Atomic Bombs', *The Pantagraph*, 31 July 2016. (https://www.newspapers.com/article/the-pantagraph-solar-storm-carried-the-e/34129633/)
5. Hodgson, R. 'On a Curious Appearance Seen in the Sun', *Monthly Notices of the Royal Astronomical Society*, 20 (1859), pp. 15–16. (https://adsabs.harvard.edu/full/1859MNRAS..20...15H/0000015.000.html)
6. Odenwald, S. F. and Green, J L. 'Bracing the Satellite Infrastructure for a Solar Superstorm', *Scientific American*, 1 August 2008. (https://www.scientificamerican.com/article/bracing-for-a-solar-superstorm/)
7. Cárdenas, Freddy Moreno, Sánchez, Sergio Cristancho, and Domínguez, Santiago Vargas. 'The Grand Aurorae Borealis Seen in Colombia in 1859', *arXiv*, 26 August 2015. (https://doi.org/10.48550/arXiv.1508.06365)
8. 'The Great Aurora of 1859', *Daily News*, 8 October 1909. (http://nla.gov.au/nla.news-article77351480)

9. Green, J. L. et al. 'Eyewitness Reports of the Great Auroral Storm of 1859', *Advances in Space Research*, 5 August 2005. (https://ntrs.nasa.gov/api/citations/20050210157/downloads/20050210157.pdf)
10. Clark, S. *The Sun Kings: The Unexpected Tragedy of Richard Carrington and the Tale of How Modern Astronomy Began*, Princeton University Press, 2019.
11. Levere, Trevor H. 'Sabine: Sir Edward', in *Dictionary of Canadian Biography*, University of Toronto/Université Laval, 1982. (https://www.biographi.ca/en/bio/5819?revision_id=33284)
12. Brehm, Nicolas et al. 'Tree-Rings Reveal Two Strong Solar Proton Events in 7176 and 5259 BC', *Nature Communications*, 13 (7 March 2022), p. 1196. (https://doi.org/10.1038/s41467-022-28804-9)
13. Boteler, D. H. 'A 21st Century View of the March 1989 Magnetic Storm', *Space Weather*, 17, (10) (2019), pp. 1427–41. (https://doi.org/10.1029/2019SW002278)
14. 'Hinkley Connection news: Supergrid Transformer Arrives in Sandford', National Grid, 19 April 2021. (https://www.nationalgrid.com/electricity-transmission/hinkley-connection/news/supergrid-transformer-arrives-sandford)
15. 'Gannon Geomagnetic Storm: Event Response Summary and Lessons Learnt', Transpower New Zealand Limited, September 2024. (https://static.transpower.co.nz/public/bulk-upload/documents/Event%204457%20-%20Gannon%20geomagnetic%20storm%20response%20summary%20and%20lessons%20learnt.pdf?VersionId=me4cehwLgGVV3f7V2ha0JafZhmikj1om)
16. Schulz, K. 'What a Major Solar Storm Could Do to Our Planet', *New Yorker*, 26 February 2024. (https://www.newyorker.com/magazine/2024/03/04/what-a-major-solar-storm-could-do-to-our-planet)
17. Andrews, R. G. 'Solar Storm Destroys 40 New SpaceX Satellites in Orbit', *New York Times*, 9 February 2022. (https://www.nytimes.com/2022/02/09/science/spacex-satellites-storm.html)
18. Knipp, D. J. et al. 'The May 1967 Great Storm and Radio Disruption Event: Extreme Space Weather and Extraordinary Responses', *Space Weather*, 14, (9) (2016), pp. 614–33. (https://doi.org/10.1002/2016SW001423)
19. National Research Council. 'Severe Space Weather Events: Understanding Societal and Economic Impacts, A Workshop Report', Washington, DC, The National Academies Press, 2008 (https://doi.org/10.17226/12507)

20. 'Transmission System Planned Performance for Geomagnetic Disturbance Events: Implementation Guidance for Reliability Standard TPL-007-4', North American Electric Reliability Corporation, November 2019. (https://www.nerc.com/pa/Stand/Project201901ModificationstoTPL0073/Draft%20Implementation%20Guidance_TPL-007_QR.pdf)
21. Ribel, M. 'Here Comes the Sun – to End Civilization', *Wired*, 21 June 2022. (https://www.wired.com/story/sun-storm-end-civilization/)
22. Skibell, A. 'This Russia-Linked Hack Is Worse than We Knew', Politico, 14 February 2023. (https://www.politico.com/newsletters/power-switch/2023/02/14/this-russia-linked-hack-worse-than-we-knew-00082755)
23. Jacobsen, *A. Nuclear War: A Scenario*, New York, Dutton, 2024, pp. 266–7.
24. Kirmani, Ahmad R. et al. 'Unraveling Radiation Damage and Healing Mechanisms in Halide Perovskites Using Energy-Tuned Dual Irradiation Dosing', *Nature Communications*, 15 (1) (24 January 2024), p. 696. (https://doi.org/10.1038/s41467-024-44876-1)
25. Ord, T. *The Precipice: Existential Risk and the Future of Humanity*, London, Hachette, 2020.
26. Hecht, J. 'Moving the Earth: A Planetary Survival Guide', *New Scientist*, 20 October 2008. (https://www.newscientist.com/article/dn14983-moving-the-earth-a-planetary-survival-guide/)

CHAPTER 3: STOPPERING A SUPERVOLCANO

1. Koeberl, C. 'Craters on the Moon From Galileo to Wegener: A Short History of the Impact Hypothesis, and Implications for the Study of Terrestrial Impact Craters', *Earth, Moon, and Planets*, 85 (1 January 1999), pp. 209–24. (https://doi.org/10.1023/A:1017020032451)
2. 'VEI: How Do You Rate the Strength of an Eruption?', Volcano Active Foundation, 19 February 2019. (https://volcanofoundation.org/vei-como-calificar-la-potencia-de-una-erupcion/)
3. Gislason, S. R. et al. 'Characterization of Eyjafjallajökull Volcanic Ash Particles and a Protocol for Rapid Risk Assessment', *Proceedings of the National Academy of Sciences*, 108 (18) (3 May 2011), pp. 7307–12. (https://doi.org/10.1073/pnas.1015053108)

4. 'Reducing the Economic Damage from Volcanic Ash Clouds', The Russell Group, 13 February 2025. (https://russellgroup.ac.uk/policy/case-studies/reducing-the-economic-damage-from-volcanic-ash-clouds/)
5. Gudmundsson, M. T. et al. 'Ash Generation and Distribution from the April-May 2010 Eruption of Eyjafjallajökull, Iceland', *Scientific Reports*, 2 (1) (14 August 2012), p. 572. (https://doi.org/10.1038/srep00572)
6. Osipov, S. et al. 'The Toba Supervolcano Eruption Caused Severe Tropical Stratospheric Ozone Depletion', *Communications Earth & Environment*, 2 (1) (12 April 2021), pp. 1–7. (https://www.nature.com/articles/s43247-021-00141-7)
7. Bradshaw Foundation. 'Origins – Exploring the Fossil Record', Bradshaw Foundation, accessed 15 February 2025. (https://www.bradshawfoundation.com/origins/index.php)
8. Smith, E. I. et al. 'Humans Thrived in South Africa through the Toba Eruption about 74,000 Years Ago', *Nature*, 555 (7697) (March 2018), pp. 511–15. (https://doi.org/10.1038/nature25967)
9. Rougier, J. et al. 'The Global Magnitude–Frequency Relationship for Large Explosive Volcanic Eruptions', *Earth and Planetary Science Letters*, 482 (15 January 2018), pp. 621–29. (https://doi.org/10.1016/j.epsl.2017.11.015)
10. Patel, K. 'What are Supervolcanoes and Should We Really Be Worried About Them?', *Washington Post*, 10 November 2023. (https://www.washingtonpost.com/climate-environment/2023/11/10/supervolcanoes-explained-eruptions-history/)
11. Wilcox, B. H. et al. 'Defending Human Civilization from Supervolcanic Eruptions', California Institute of Technology, 2015. (https://scienceandtechnology.jpl.nasa.gov/sites/default/files/documents/DefendingCivilizationFromSupervolcanos20151015.pdf)
12. 'High Plains Regional Ground-Water Study', US Department of the Interior and US Geological Survey, August 2000. (https://pubs.usgs.gov/fs/2000/0091/report.pdf)
13. 'Volcanoes', British Geological Survey, accessed 15 February 2025. (https://www.bgs.ac.uk/discovering-geology/earth-hazards/volcanoes/)
14. Bahr, L. 'A Celebrity Volcanologist Couple Spotlighted in New Doc', Associated Press, 5 July 2022. (https://apnews.com/article/entertainment-fires-volcanoes-werner-herzog-20d02f19904e1b326148add52b41721a)

15. Keller, J. 'Memorial for Katja and Maurice Krafft', *Bulletin of Volcanology*, 54 (7) (1 September 1992), pp. 613–14. (https://doi.org/10.1007/BF00569946)
16. Calderazzo, J. 'Fire in the Earth, Fire in the Soul: The Final Moments of Maurice and Katia Krafft', *Interdisciplinary Studies in Literature and Environment*, 4 (2) (1997), pp. 71–77. (https://www.jstor.org/stable/44085498)
17. Newhall, C. and Solidum, R. U. 'Volcanic Hazard Communication at Pinatubo from 1991 to 2015', in Fearnley, C. J. et al (eds). *Observing the Volcano World: Volcano Crisis Communication*, Cham, Springer International Publishing, 2018. (https://doi.org/10.1007/11157_2016_43)
18. Spence, R. et al. 'Building Damage Caused by the Mount Pinatubo Eruption of June 15, 1991', *Fire and Mud: Eruptions and Lahars of Mount Pinatubo, Philippines*, 1 January 1996, pp. 1055–61. (https://pubs.usgs.gov/pinatubo/spence/)
19. 'Impacts & Mitigation – Pinatubo 1991', US Geological Survey Volcanic Ashfall Impacts Working Group, accessed 15 February 2025. (https://volcanoes.usgs.gov/volcanic_ash/pinatubo_1991.html)
20. Newhall and Solidum, 2018.
21. Yirka, B. 'Study Shows People Died from Body Fluid Vaporization Due to Pyroclastic Flows from Vesuvius', Phys.org, 10 October 2018. (https://phys.org/news/2018-10-people-died-body-fluid-vaporization.html)
22. 'US Volcano Expert's Body Found in Japan', *Los Angeles Times*, 7 June 1991. (https://www.latimes.com/archives/la-xpm-1991-06-07-mn-126-story.html)
23. Sandberg, A. 'Blueberry Earth', *arXiv*, 27 July 2018. (https://doi.org/10.48550/arXiv.1807.10553)
24. de la Garza, A. 'A Controversial Technology is Creating an Unprecedented Rift Among Climate Scientists', *TIME*, 17 March 2023. (https://time.com/6264143/geoengineering-climate-scientists-divided/)
25. Hamilton, C. *Earth Masters: Playing God with the Climate*, Sydney, Allen and Unwin, 2013.
26. Talati, S. 'A New Era of Policy in Solar Geoengineering', Kleinman Center for Energy Policy, 12 January 2024. (https://kleinmanenergy.upenn.edu/research/publications/a-new-era-of-policy-in-solar-geoengineering/)
27. Janssens, R. V. A. *'What Future for Japan?': US Wartime Planning for the Postwar Era, 1942–1945*, Leiden, Rodolpi, 1995.

28. Longman, J. et al. 'Viability of Greenhouse Gas Removal via Artificial Addition of Volcanic Ash to the Ocean', *Anthropocene*, 32 (1 December 2020), 100264. (https://doi.org/10.1016/j.ancene.2020.100264)
29. Dubrow, A. 'A Volcanic Idea to Reverse Climate Change', US National Science Foundation, 19 August 2011. (https://www.nsf.gov/news/volcanic-idea-reverse-climate-change)
30. The Office of Communications and Publishing. 'New England's 1816 "Mackerel Year", Volcanoes and Climate Change Today', US Geological Survey, 18 January 2017. (https://www.usgs.gov/news/featured-story/new-englands-1816-mackerel-year-volcanoes-and-climate-change-today)
31. Marshall, A. et al. 'The Tambora–Frankenstein Myth: The Monster Inspired', *Humanities Bulletin*, 2 (2019), pp. 217–35. (https://core.ac.uk/reader/266992850)
32. Marshall, A. 'Did a Volcanic Eruption in Indonesia Really Lead to the Creation of Frankenstein?', The Conversation, 24 January 2020. (http://theconversation.com/did-a-volcanic-eruption-in-indonesia-really-lead-to-the-creation-of-frankenstein-130212)
33. Caldero, R. 'India's Strategic Edge: Controlling China's Oil Lifeline', *Rio Times*, 22 November 2024. (https://www.riotimesonline.com/indias-strategic-edge-controlling-chinas-oil-lifeline/)
34. Sari, M. N. et al. 'Indonesian Geostrategy on International Maritime Transportation in the Malacca Strait', *International Journal of Progressive Sciences and Technologies*, 37 (2) (2023), pp. 372–81. (https://ijpsat.org/index.php/ijpsat/article/download/5171/3224)
35. *Making Trade Work for the Environment, Prosperity and Resilience*, World Trade Organisation and UN Environment, 2018. (https://doi.org/10.30875/f14f8c90-en)
36. 'Country Analysis Brief: World Oil Transit Chokepoints', Washington, DC, US Department of Energy, 2024. (https://www.eia.gov/international/content/analysis/special_topics/World_Oil_Transit_Chokepoints/wotc.pdf)
37. Bailey, R. and Wellesley, L. *Chokepoints and Vulnerabilities in Global Food Trade*, London, Chatham House, 2017. (https://www.chathamhouse.org/sites/default/files/publications/research/2017-06-27-chokepoints-vulnerabilities-global-food-trade-bailey-wellesley-final.pdf)

38. Mani, L. et al. 'Global Catastrophic Risk from Lower Magnitude Volcanic Eruptions', *Nature Communications*, 12 (1) (6 August 2021), 4756. (https://doi.org/10.1038/s41467-021-25021-8)
39. 'Indonesia's Sinabung Volcano Spews High Column of Ash', Associated Press, 2 March 2021. (https://ca.news.yahoo.com/indonesias-sinabung-volcano-spews-high-042241572.html)
40. Roszandi, D. 'Merapi Volcano Eruption Continues, Emitting Lava Flows and Fumes in Central Java', Anadolu Agency, accessed 15 February 2025. (https://www.aa.com.tr/en/pg/photo-gallery/merapi-volcano-eruption-continues-emitting-lava-flows-and-fumes-in-central-java)
41. 'Piracy Disappears After Tsunami', NBC News, 9 February 2009. (https://www.nbcnews.com/id/wbna6937248)
42. Koh, F. 'Budget Debate: $90m to be Spent on Singapore's First Top-Level Biosafety Lab, to be Pperational by 2025', *Straits Times*, 1 March 2021. (https://www.straitstimes.com/singapore/politics/90-million-to-be-spent-on-singapores-first-top-level-biosafety-lab-to-be)
43. Gambrell, J. '3 Red Sea Data Cables Cut as Houthis Launch More Attacks in the Vital Waterway', Associated Press, 5 March 2024. (https://apnews.com/article/red-sea-undersea-cables-yemen-houthi-rebels-attacks-b53051f61a41bd6b357860bbf0b0860a)
44. Abraham, T. K. 'Thailand Takes $28 Billion Malacca Strait Bypass Plan to US', Bloomberg, 14 November 2023. (https://www.bloomberg.com/news/articles/2023-11-14/thailand-pitches-28-billion-malacca-strait-bypass-project-to-us)
45. Babones, S. 'The Next Front in the India-China Conflict Could Be a Thai Canal', *Foreign Policy*, 1 September 2020. (https://foreignpolicy.com/2020/09/01/china-india-conflict-thai-kra-canal/)
46. 'Volcanic Coal-Burning in Siberia Led to Climate Change 252 Million Years Ago', US National Science Foundation, 18 June 2020. (https://www.nsf.gov/news/volcanic-coal-burning-siberia-led-climate-change)

CHAPTER 4: 'WE WILL DIE, BUT WE WILL SINK THEM ALL!'

1. Brian Cathcart, *The Fly in the Cathedral: How a Group of Cambridge Scientists Won the Race to Split the Atom*, London, Viking, 2004.
2. Spence, R. 'Otto Hahn – Nobel Prize-Winning German Chemist', *Encyclopedia Britannica*, 24 July 2024. (https://www.britannica.com/biography/Otto-Hahn)

3. Lipscombe, T. 'Einstein Feared a Nazi Atom Bomb – But Immigrants Made Sure the US Got There First', *TIME*, 2 August 2019. (https://time.com/5641891/einstein-szilard-letter/)
4. Einstein, A. 'Einstein–Szilard Letter', *Atomic Heritage Foundation*, 2 August 1939. (https://ahf.nuclearmuseum.org/ahf/key-documents/einstein-szilard-letter/)
5. 'MAUD Committee Report', *Atomic Heritage Foundation*, 1941. (https://ahf.nuclearmuseum.org/ahf/key-documents/maud-committee-report/)
6. Kleinman, Z. and Vallance, C. 'AI "godfather" Geoffrey Hinton Warns of Dangers as He Quits Google', BBC News, 2 May 2023. (https://www.bbc.com/news/world-us-canada-65452940)
7. Lewis, D. 'The World's First Nuclear Reactor Was Built in a Squash Court', *Smithsonian Magazine*, 25 November 2015. (https://www.smithsonianmag.com/smart-news/worlds-first-nuclear-reactor-was-built-squash-court-180957390/)
8. Long, T. 'July 16, 1945: Trinity Blast Opens Atomic Age', *Wired*, 16 July 2008. (https://www.wired.com/2008/07/dayintech-0716-2/)
9. Quote from Long, 2008.
10. Barnett, L. 'J. Robert Oppenheimer', *LIFE*, 10 October 1949. (https://books.google.com/books?id=GVIEAAAAMBAJ&dq=oppenheimer+%22if+the+radiance+of+a+thousand+suns+were+to+%22&pg=PA133#v=onepage&q=oppenheimer%20%22if%20the%20radiance%20of%20a%20thousand%20suns%20were%20to%20%22&f=false)
11. 'Little Boy – Hiroshima – August 6, 1945', Navy Museum of the US Navy, accessed 15 February 2025. (http://public1.nhhcaws.local/content/history/museums/nmusn/explore/photography/wwii/wwii-pacific/bombardment-japan/bombs-atomic/little-boy-hiroshima.html)
12. 'The Smithsonian and the Enola Gay', in AFA's Enola Gay Controversy Archive Collection, Air Force Association, accessed 15 February 2025. (https://www.airandspaceforces.com/PDF/SiteCollectionDocuments/Enola%20Gay%20Archive/EG_crew.pdf)
13. This whole paragraph is informed by Schlosser, E. *Command and Control: Nuclear Weapons, the Damascus Accident, and the Illusion of Safety*, New York, Penguin, 2013, pp. 50–1.
14. Details in this passage are largely from 'Manhattan Project: The Atomic Bombing of Hiroshima, August 6, 1945', *The Manhattan Project: An Interactive History*, US Department of Energy, accessed

15 February 2025. (https://www.osti.gov/opennet/manhattan-project-history/Events/1945/hiroshima.htm)

15. '"We Learned to Step over the Dead": Hiroshima Survivor & Anti-Nuclear Activist Recalls US Bombing', *Democracy Now!*, 27 May 2016. (http://www.democracynow.org/2016/5/27/we_learned_to_step_over_the)
16. Hsu, W-L. et al. 'The Incidence of Leukemia, Lymphoma, and Multiple Myeloma among Atomic Bomb Survivors: 1950–2001', *Radiation Research*, 179 (3) (March 2013), 10.1667/RR2892.1. (https://doi.org/10.1667/RR2892.1)
17. Wellerstein, A. 'Counting the Dead at Hiroshima and Nagasaki', Bulletin of the Atomic Scientists, 4 August 2020. (https://thebulletin.org/2020/08/counting-the-dead-at-hiroshima-and-nagasaki/)
18. 'Foreign Relations of the United States: Diplomatic Papers, 1945, the British Commonwealth, the Far East, Volume vi', Office of the Historian, Foreign Service Institute United States Department of State, 6 August 1945. (https://history.state.gov/historicaldocuments/frus1945v06/d401)
19. 'R.A.F's "GRAND SLAM"', Imperial War Museum, accessed 15 February 2025. (https://www.iwm.org.uk/collections/item/object/205454520)
20. Hughes, T. A. and Royde-Smith, J. G. 'World War II – Hiroshima, Nagasaki, Atomic Bombs', Encyclopedia Britannica, accessed 14 February 2025. (https://www.britannica.com/event/World-War-II/Hiroshima-and-Nagasaki)
21. Kumlertsakul, P. 'Soviet-Japan and the Termination of the Second World War', The National Archives, 2 September 2020. (https://blog.nationalarchives.gov.uk/soviet-japan-and-the-termination-of-the-second-world-war/)
22. Wilson, W. 'The Bomb Didn't Beat Japan. Stalin Did', *Foreign Policy*, 30 May 2013. (https://foreignpolicy.com/2013/05/30/the-bomb-didnt-beat-japan-stalin-did/)
23. Rees, M. *Our Final Hour: A Scientist's Warning – How Terror, Error, and Environmental Disaster Threaten Humankind's Future in This Century – On Earth and Beyond*, London, Basic Books, 2003, p. 39.
24. 'Detection of the First Soviet Nuclear Test, September 1949', National Security Archive, 9 September 2019. (https://nsarchive.gwu.edu/briefing-book/nuclear-vault/2019-09-09/detection-first-soviet-nuclear-test-september-1949)

25. Mills, C. 'Nuclear Weapons at a Glance: United Kingdom', House of Commons Library, 1 August 2024. (https://commonslibrary.parliament.uk/research-briefings/cbp-9077/)
26. 'Israel's Nuclear Weapon Capability: An Overview', Wisconsin Project on Nuclear Arms Control, 29 April 2015. (https://web.archive.org/web/20150429192508/http:/www.wisconsinproject.org/countries/israel/nuke.html)
27. 'Estimated Number of Nuclear Warheads per Country in Each Year from 1945 to 2023', Statista, 4 July 2024. (https://www.statista.com/statistics/1071026/nuclear-warheads-per-country-historical-development/)
28. Chan, M. 'What Is the Difference Between a Hydrogen Bomb and an Atomic Bomb?', *TIME*, 22 September 2017. (https://time.com/4954082/hydrogen-bomb-atomic-bomb/)
29. 'The Chairman of the General Advisory Committee (Oppenheimer) to the Chairman of the United States Atomic Energy Commission (Lilienthal)', Office of the Historian, Foreign Service Institute United States Department of State, 30 October 1949. (https://history.state.gov/historicaldocuments/frus1949v01/d211)
30. Jacobsen, A. *Nuclear War: A Scenario*, New York, Dutton, 2004, p. 24.
31. Kratz, J. 'Brush with Catastrophe: The Day the US Almost Nuked Itself', US National Archives Pieces of History, 22 January 2021. (https://prologue.blogs.archives.gov/2021/01/22/brush-with-catastrophe-the-day-the-u-s-almost-nuked-itself/)
32. 'The Cuban Missile Crisis', US Department of State, 27 May 2010. (https://web.archive.org/web/20100527171609/http:/future.state.gov/educators/slideshow/cuba/cuba2.html)
33. 'The Cuban Missile Crisis at 50: What's Your Story?', *The Nuclear Threat Initiative*, 15 October 2012. (https://www.nti.org/news/cuban-missile-crisis-50-whats-your-story/)
34. Kolbert, E. 'The Day Nuclear War Almost Broke Out', *New Yorker*, 5 October 2020. (https://www.newyorker.com/magazine/2020/10/12/the-day-nuclear-war-almost-broke-out)
35. Burr, W. and Blanton, T. S. (eds), 'The Submarines of October: US and Soviet Naval Enxcounters During the Cuban Missile Crisis', in National Security Archive Electronic Briefing Book No. 75, 2002. (https://nsarchive2.gwu.edu/NSAEBB/NSAEBB75/)
36. Ibid.
37. Hoffman, D. '"I Had a Funny Feeling in My Gut"', *Washington Post*, 10 February 1999. (https://www.washingtonpost.com/archive/

politics/1999/02/10/i-had-a-funny-feeling-in-my-gut/97fb4af6-2a54-4805-8ec0-115a13f33493/)

38. Ibid.

39. Chan, S. 'Stanislav Petrov, Soviet Officer Who Helped Avert Nuclear War, is Dead at 77', *New York Times*, 18 September 2017. (https://www.nytimes.com/2017/09/18/world/europe/stanislav-petrov-nuclear-war-dead.html)

40. Aksenov, P. 'Stanislav Petrov: The Man Who May Have Saved the World', BBC News, 26 September 2013. (https://www.bbc.com/news/world-europe-24280831)

41. Xia, L. et al. 'Global Food Insecurity and Famine from Reduced Crop, Marine Fishery and Livestock Production Due to Climate Disruption from Nuclear War Soot Injection', *Nature Food*, 3 (8) (August 2022), pp. 586–96. (https://doi.org/10.1038/s43016-022-00573-0)

42. Ibid.

43. Khrushchev, N. S. 'The Living Will Envy the Dead', in Platt, S. (ed.), *Respectfully Quoted: A Dictionary of Quotations*, Bartleby, 2022.

44. US edition, p. 115.

45. McLaughlin, T. 'Defence Agency Stopped Delivery on Raytheon Warheads', *Boston Business Journal*, 25 March 2011, quoted in Jacobsen, 2024, p. 73.

46. Jacobsen, 2024, p. 73.

47. 'The Ballistic Missile Defense System', *Missile Defence Review*, accessed 15 February 2025. (https://www.defense.gov/Portals/1/Interactive/2018/11-2019-Missile-Defense-Review/MDR-BMDS-Factsheet-UPDATED.pdf)

48. Fink, A. L. 'Russia's Nuclear Weapons', Congressional Research Service, 22 January 2025. (https://crsreports.congress.gov/product/pdf/IF/IF12672)

49. 'Ballistic Missile Defense System (BMDS)', Office of the Director, Operational Test & Evaluation, accessed 15 February 2025. (https://www.dote.osd.mil/Portals/97/pub/reports/FY2012/bmds/2012bmds.pdf?ver=2019-08-22-111711-520)

50. 'Advanced Future Military Laser Achieves UK First', Defence Science and Technology Laboratory and Ministry of Defence, 21 March 2024. (https://www.gov.uk/government/news/advanced-future-military-laser-achieves-uk-first)

51. Goettemoeller, R. *Negotiating the New START Treaty*, Amherst, Cambria Press, 2021, p. xv.

52. 'Biography: Rose Gottemoeller', US Department of State, 17 April 2009. (https://web.archive.org/web/20090417061214/http:/www.state.gov/r/pa/ei/biog/121630.htm)
53. Gottemoeller, 2021, p. 26.
54. Ibid. pp. 22–3.
55. Ibid. p. 125.
56. Ibid. pp. 145–6
57. 'New START Treaty', The Nuclear Threat Initiative, accessed 15 February 2025. (https://www.nti.org/education-center/treaties-and-regimes/treaty-between-the-united-states-of-america-and-the-russian-federation-on-measures-for-the-further-reduction-and-limitation-of-strategic-offensive-arms/)
58. Greenemeier, L. 'Russia's New Nukes Are Similar to a Risky Project the US Abandoned', *Scientific American*, 2 March 2018. (https://www.scientificamerican.com/article/russias-new-nukes-are-similar-to-a-risky-project-the-u-s-abandoned/)
59. Anderson, E. and Sheftalovich, Z. 'Putin Wins Right to Extend His Rule until 2036 in Landslide Vote', Politico, 1 July 2020. (https://www.politico.eu/article/vladimir-putin-wins-right-to-extend-his-russia-rule-until-2036-in-landslide-vote/)
60. Hayden, J. 'Russia's Medvedev Threatens to Turn Kyiv into "Giant Melted Spot"', Politico, 14 September 2024. (https://www.politico.eu/article/russia-dmitry-medvedev-ukraine-kyiv-war-giant-melted-spot-threat/)
61. 'Kremlin Says Russia Will Appoint a New Ambassador to the United States', Reuters, 7 October 2024. (https://www.reuters.com/world/europe/kremlin-says-russia-will-appoint-new-ambassador-united-states-2024-10-07/)
62. Woodward, B. *War*, New York, Simon & Schuster, 2024
63. Gallagher, S. 'Launch Code for US Nukes Was 00000000 for 20 Years', Ars Technica, 3 December 2013. (https://arstechnica.com/tech-policy/2013/12/launch-code-for-us-nukes-was-00000000-for-20-years/)

CHAPTER 5: FEEDING THE BENIGHTED

1. Alvarez, L. W. et al. 'Extraterrestrial Cause for the Cretaceous-Tertiary Extinction', *Science*, 208 (4448) (6 June 1980), pp. 1095–1108. (https://doi.org/10.1126/science.208.4448.1095)
2. Vajda, V. and McLoughlin, S. 'Fungal Proliferation at the Cretaceous-Tertiary Boundary', *Science*, 303 (5663)

(5 March 2004), p. 1489. (https://doi.org/10.1126/science.1093807)

3. Wiblin, R. and Harris, K. '#117 – David Denkenberger on Using Paper Mills and Seaweed to Feed Everyone in a Catastrophe, Ft Sahil Shah', 80,000 Hours, 29 November 2021. (https://80000hours.org/podcast/episodes/david-denkenberger-sahil-shah-using-paper-mills-and-seaweed-in-catastrophes/)
4. Bostrom, N. 'Existential Risk Prevention as Global Priority', *Global Policy*, 4 (1) (2013), pp. 15–31. (https://doi.org/10.1111/1758-5899.12002)
5. Aizen, M. A. et al. 'How Much Does Agriculture Depend on Pollinators? Lessons from Long-Term Trends in Crop Production', *Annals of Botany*, 103 (9) (June 2009), pp. 1579–88. (https://doi.org/10.1093/aob/mcp076)
6. Ibid.
7. Jehn, F. U. et al. 'Seaweed as a Resilient Food Solution After a Nuclear War', *Earth's Future*, 12 (1) (January 2024), e2023EF003710. (https://doi.org/10.1029/2023EF003710)
8. Throup, J. et al. 'Rapid Repurposing of Pulp and Paper Mills, Biorefineries, and Breweries for Lignocellulosic Sugar Production in Global Food Catastrophes', Allfed, 2021. (https://allfed.info/images/pdfs/rapid_repurposing.pdf)
9. Denkenberger, D. C. and Blair, R. W. 'Interventions That May Prevent or Mollify Supervolcanic Eruptions', *Futures*, 102 (1 September 2018) pp. 51–62. (https://doi.org/10.1016/j.futures.2018.01.002)
10. Dathan, M. 'Are You Ready for a Crisis? Government Says Stock up for Three Days', *The Times*, 21 May 2024. (https://www.thetimes.com/uk/politics/article/stock-up-for-three-days-in-case-of-national-crisis-official-advice-0jp5t7s9c)
11. Avery, D. T. 'Winning the Food Race', *The Brown Journal of World Affairs*, 18 (1) (2011), pp. 107–18. (https://www.jstor.org/stable/24590780)
12. 'President Bush Presents Congressional Gold Medal to Dr. Norman Borlaug', The White House, 17 July 2007. (https://georgewbush-whitehouse.archives.gov/news/releases/2007/07/20070717-1.html)
13. Dathan, 2024.
14. 'President Bush Presents Congressional Gold Medal to Dr Norman Borlaug', 2007.
15. Ough, T. 'Why Britain Could Fare Worse than Other Nations in a Global Catastrophe', *Telegraph*, 20 April 2022. (https://www.

telegraph.co.uk/news/2022/04/20/britain-could-fare-worse-nations-global-catastrophe/)

16. Hu, W. et al. 'Genomic Inference of a Severe Human Bottleneck during the Early to Middle Pleistocene Transition', *Science*, 381 (6661) (September 2023), pp. 979–84. (https://doi.org/10.1126/science.abq7487)
17. Edmunds, M. 'Antikythera mechanism', *Encyclopaedia Britannica*, accessed 15 February 2025. (https://www.britannica.com/topic/Antikythera-mechanism)

CHAPTER 6: TEARING OFF THE HAIRSHIRT

1. Crawford, E. 'Svante Arrhenius', *Encyclopedia Britannica*, accessed 15 February 2025. (https://www.britannica.com/biography/Svante-Arrhenius)
2. Callendar, G. S. 'The Artificial Production of Carbon Dioxide and Its Influence on Temperature', *Quarterly Journal of the Royal Meteorological Society*, 64 (275) (1938), pp. 223–40. (https://doi.org/10.1002/qj.49706427503)
3. Mann, C. C. 'Meet the Amateur Scientist Who Discovered Climate Change', Wired, 23 January 2018. (https://www.wired.com/story/meet-the-amateur-scientist-who-discovered-climate-change/)
4. 'Climate Change Occurring Ten Times Faster than at Any Time in Past 65 Million Years', ScienceDaily, 1 August 2013. (https://www.sciencedaily.com/releases/2013/08/130801142420.htm)
5. Roser, M. 'Extreme Poverty: How Far Have We Come, and How Far Do We Still Have to Go?', Our World in Data, 22 November 2021. (https://ourworldindata.org/extreme-poverty-in-brief)
6. MacAskill, W. *What We Owe the Future*, New York, Basic Books, 2023, pp. 140–2.
7. 'Greta Thunberg to World Leaders: "How Dare You? You Have Stolen My Dreams and My Childhood"', *Guardian News*, YouTube, 2019. (https://www.youtube.com/watch?v=TMrtLsQbaok)
8. Brulle, R. '30 Years Ago Global Warming Became Front-Page News – and Both Republicans and Democrats Took It Seriously', The Conversation, 19 June 2018. (http://theconversation.com/30-years-ago-global-warming-became-front-page-news-and-both-republicans-and-democrats-took-it-seriously-97658)
9. Shabecoff, P. 'Global Warming Has Begun, Expert Tells Senate', *New York Times*, 24 June 1988. (https://www.nytimes.

com/1988/06/24/us/global-warming-has-begun-expert-tells-senate.html)

10. Wiblin, R. and Harris, K. '#148 – Johannes Ackva on Unfashionable Climate Interventions That Work, and Fashionable Ones That Don't', 80,000 Hours, 3 April 2023. (https://80000hours.org/podcast/episodes/johannes-ackva-unfashionable-climate-interventions/)
11. Intergovernmental Panel on Climate Change, *Climate Change 2021 – The Physical Science Basis: Working Group I Contribution to the Sixth Assessment Report of the Intergovernmental Panel on Climate Change*, Cambridge University Press, 2023. (https://doi.org/10.1017/9781009157896).
12. Intergovernmental Panel on Climate Change, *Climate Change 2023: Synthesis Report*, 2023. (https://www.ipcc.ch/report/ar6/syr/downloads/report/IPCC_AR6_SYR_SPM.pdf)
13. Ballard, J. 'Nearly Two in Five Americans Say It's Likely Climate Change Will Make the Earth Uninhabitable', YouGov, 22 April 2022. (https://today.yougov.com/society/articles/42262-americans-say-climate-change-earth-uninhabitable)
14. Hausfather, Z. 'State of the Climate: 2023 Smashes Records for Surface Temperature and Ocean Heat', Carbon Brief, 12 January 2024. (https://www.carbonbrief.org/state-of-the-climate-2023-smashes-records-for-surface-temperature-and-ocean-heat/)
15. Poynting, M. et al. '2024 First Year to Pass 1.5C Global Warming Limit', BBC News, 10 January 2025. (https://www.bbc.com/news/articles/cd7575x8yq5o)
16. Ritchie, H. 'Solar Panel Prices Have Fallen by Around 20% Every Time Global Capacity Doubled', Our World in Data, 13 June 2024. (https://ourworldindata.org/data-insights/solar-panel-prices-have-fallen-by-around-20-every-time-global-capacity-doubled)
17. Evans, S. and Viisainen, V. 'Analysis: Fossil Fuels Fall to Record-Low 2.4% of British Electricity', Carbon Brief, 24 April 2024. (https://www.carbonbrief.org/analysis-fossil-fuels-fall-to-record-low-2-4-of-british-electricity/)
18. Ibid.
19. 'California's Grid Keeps Setting New Clean Energy Records', Governor of California, 19 April 2024. (https://www.gov.ca.gov/2024/04/19/californias-grid-keeps-setting-new-clean-energy-records/)
20. Southwood, B. et al. 'Foundations: Why Britain Has Stagnated', UK Foundations, accessed 15 February 2025. (https://ukfoundations.co)

21. Fraser, T. 'A Nuclear Kettle of Fish at Hinkley Point C', Politics Home, 15 October 2024. (https://www.politicshome.com/thehouse/article/fish-disco-hinkley-point-c-nuclear-energy)
22. Dobrin, N. 'Kim Kardashian Traveled to Paris from LA Just for Cheesecake?', Snopes, 13 June 2024. (https://www.snopes.com//fact-check/kim-kardashian-paris-cheesecake/)
23. 'Reactor Core', US Nuclear Regulatory Commission, 9 March 2021. (https://www.nrc.gov/reading-rm/basic-ref/glossary/reactor-core.html)
24. 'INSAG-7 The Chernobyl Accident: In Updating of INSAG-1', Vienna, International Atomic Energy Agency, 1992. (https://www-pub.iaea.org/MTCD/publications/PDF/Pub913e_web.pdf)
25. 'Chernobyl', excerpt from Rhodes, R. *Nuclear Renewal*, New York, Penguin, 1993.
26. Ritchie, H. 'What Was the Death Toll from Chernobyl and Fukushima?', Our World in Data, 24 July 2017. (https://ourworldindata.org/what-was-the-death-toll-from-chernobyl-and-fukushima)
27. 'Air Pollution', World Health Organisation, accessed 15 February 2025. (https://www.who.int/health-topics/air-pollution)
28. Ritchie, H. 'What Are the Safest and Cleanest Sources of Energy?', Our World in Data, 10 February 2020. (https://ourworldindata.org/safest-sources-of-energy)
29. Lelieveld, J. et al. 'Air Pollution Deaths Attributable to Fossil Fuels: Observational and Modelling Study', *British Medical Journal*, 383 (29 November 2023), e077784. (https://doi.org/10.1136/bmj-2023-077784)
30. 'Debate', *Western Daily Press*, 19 February 1981. (https://www.newspapers.com/article/western-daily-press-debate/148176282/)
31. Crellin, F. et al. 'EDF's UK Hinkley Point Nuclear Plant Start Date Delayed Again, Costs Mount', Reuters, 25 January 2024. (https://www.reuters.com/business/energy/edfs-nuclear-project-britain-pushed-back-2029-may-cost-up-34-bln-2024-01-23/)
32. Dumitriu, S. 'Why Britain is Building the World's Most Expensive Nuclear Plant', *Spectator*, 2 May 2024. (https://www.spectator.co.uk/article/why-britain-is-building-the-worlds-most-expensive-nuclear-plant/)
33. 'Fukushima Daiichi Accident', World Nuclear Association, 29 April 2024. (https://world-nuclear.org/information-library/safety-and-security/safety-of-plants/fukushima-daiichi-accident)

34. Yamaguchi, M. 'A Tiny Grain of Nuclear Fuel is Pulled from Ruined Japanese Nuclear Plant, in a Step toward Cleanup', Associated Press, 8 November 2024. (https://apnews.com/article/japan-nuclear-fukushima-sample-robot-decommissioning-b7dd956a86401b39f04ae839f768d29f)
35. 'Three Mile Island Accident', *Encyclopedia Britannica*, 24 January 2025. (https://www.britannica.com/event/Three-Mile-Island-accident)
36. Fisher, S. 'For Germany's Green Party, the 50-Year Dream to End Nuclear Power Ends in a Nightmare', *The Chicago Journal of Foreign Policy*, 6 June 2024. (https://www.cjfp.org/for-germanys-green-party-the-50-year-dream-to-end-nuclear-power-ends-in-a-nightmare/)
37. 'Germany Approves Bringing Coal-Fired Power Plants Back Online This Winter', Reuters, 4 October 2023. (https://www.reuters.com/business/energy/germany-approves-bringing-coal-fired-power-plants-back-online-this-winter-2023-10-04/)
38. Conca, J. 'A Conversation with Grace Stanke, Miss America 2023', NuclearNewswire, 24 August 2023. (https://www.ans.org/news/article-5193/a-conversation-with-grace-stanke-miss-america-2023/)
39. 'The NEA Small Modular Reactor Dashboard: Second Edition', Nuclear Energy Agency, 13 March 2024. (https://www.oecd-nea.org/jcms/pl_90816/the-nea-small-modular-reactor-dashboard-second-edition?details=true)
40. 'Poland Approves Construction of SMR Nuclear Units at Six Sites', Reuters, 7 December 2023. (https://www.reuters.com/sustainability/climate-energy/poland-approves-construction-smr-nuclear-units-six-sites-2023-12-07/)
41. Perera, J. 'South Korea's SMART100 SMR Approved', *Nuclear Engineering International*, 3 October 2024. (https://www.neimagazine.com/news/south-koreas-smart100-smr-approved/)
42. Pashby, T. 'Last Energy Plans Four 20MW Micro Reactors in South Wales to Support Manufacturing', *New Civil Engineer*, 15 October 2024. (https://www.newcivilengineer.com/latest/last-energy-plans-four-20mw-micro-reactors-in-south-wales-to-support-manufacturing-15-10-2024/)
43. Ough, T. 'The Cement That Could Turn Your House into a Giant Battery', BBC, 11 June 2024. (https://www.bbc.com/future/article/20240610-how-the-concrete-in-your-house-could-be-turned-into-a-battery)

44. Dewan, A. 'Scientists Just Set a Nuclear Fusion Record in a Step toward Unleashing the Limitless, Clean Energy Source', CNN, 8 February 2024. (https://www.cnn.com/2024/02/08/climate/nuclear-fusion-energy-milestone-climate/index.html)
45. Cowen, T. 'Nuclear Fusion Is Close Enough to Start Dreaming', Bloomberg, 11 November 2021. (https://www.bloomberg.com/opinion/articles/2021-11-11/nuclear-fusion-is-close-enough-to-start-dreaming-of-a-new-world)
46. Lecamwasam, T. 'Three Emerging Technologies for Low-Carbon Concrete', ClimateWorks Foundation, 30 July 2024. (https://www.climateworks.org/blog/three-emerging-technologies-for-low-carbon-concrete/)
47. Ritchie, H. et al. 'Greenhouse Gas Emissions', Our World in Data, 10 June 2020. (https://ourworldindata.org/greenhouse-gas-emissions)
48. Calma, J. 'The World's Largest Carbon Removal Plant is Here, and Bigger Ones are on the Way', The Verge, 8 May 2024. (https://www.theverge.com/2024/5/8/24151905/mammoth-carbon-removal-direct-air-capture-climeworks)
49. Nilsen, E. 'Biden Administration to Invest $1.2 Billion in Projects to Suck Carbon out of the Air', CNN, 11 August 2023. (https://www.cnn.com/2023/08/11/politics/direct-air-capture-funding-biden-administration/index.html)
50. 'Energy Fundamentals of Carbon Removal', Ben James, accessed 15 February 2025. (https://climate.benjames.io/energy-carbon-removal/)
51. Bastin, J-F. et al. 'The Global Tree Restoration Potential', *Science*, 365 (6448) (5 July 2019), pp. 76–79. (https://doi.org/10.1126/science.aax0848)
52. Alamalhodaei, A. 'Terraform Industries Converts Electricity and Air into Synthetic Natural Gas for the First Time', TechCrunch, 1 April 2024. (https://techcrunch.com/2024/04/01/terraform-industries-converted-electricity-and-air-into-synthetic-natural-gas/)
53. Ough, T. 'How to Terraform Mars, without Nukes, on a Budget', Freethink, 30 May 2023. (https://www.freethink.com/space/how-to-terraform-mars)
54. 'The Solar Revolution Turning Sunlight into Synthetic Fuel', Freethink, 29 August 2024. (https://www.freethink.com/series/the-freethink-interview/casey-handmer)
55. Ough, T. 'The Architect Fighting for Sustainability without the Greenwash', *Prospect*, 14 June 2023. (https://www.

prospectmagazine.co.uk/views/people/61728/andrew-waugh-architect-mass-timber-climate)

56. Pultarova, T. 'Breakthrough Coming? Iceland Could Get Solar Power from Space in 2030', Space.com, 24 October 2024. (https://www.space.com/iceland-space-based-solar-power)
57. Pogue, D. 'The Electric-Plane Future Is about to Take Off', CBS News, 5 September 2024. (https://www.cbsnews.com/news/electric-planes-evtol-eplanes-future-of-aviation/)
58. Crown, C. 'Some Countries Are Ending Support for EVs. Is It Too Soon?', *MIT Technology Review*, 23 September 2024. (https://www.technologyreview.com/2024/09/23/1104247/ending-ev-subsidies/)
59. Miller, S. A. 'Five Misperceptions Surrounding the Environmental Impacts of Single-Use Plastic', *Environmental Science & Technology*, 54 (22) (17 November 2020), pp. 14143–51. (https://doi.org/10.1021/acs.est.0c05295)
60. Parker, D. E. et al. 'The Impact of Mount Pinatubo on World-Wide Temperatures', *International Journal of Climatology*, 16 (5) (1996), pp. 487–97. (https://doi.org/10.1002/(SICI)1097-0088(199605)16:5<487::AID-JOC39>3.0.CO;2-J)
61. Houser, K. 'The Startup Using Balloons to Cool the Planet', Freethink, 21 September 2024. (https://www.freethink.com/energy/sai-make-sunsets)
62. Temple, J. 'Harvard Has Halted Its Long-Planned Atmospheric Geoengineering Experiment', *MIT Technology Review*, 18 March 2024. (https://www.technologyreview.com/2024/03/18/1089879/harvard-halts-its-long-planned-atmospheric-geoengineering-experiment/)
63. Henriksen, C. et al. 'Regarding SCoPEx Plans for Test Flights at the Swedish Space Corporation in Kiruna', 24 February 2021. (https://static1.squarespace.com/static/5dfb35a66f00d54ab0729b75/t/603e2167a9c0b96ffb027c8d/1614684519754/Letter+to+Scopex+Advisory+Committee+24+February.pdf)
64. Boretti, A. 'Reassessing the Cooling That Followed the 1991 Volcanic Eruption of Mt. Pinatubo', *Journal of Atmospheric and Solar-Terrestrial Physics*, 256 (1 March 2024), pp. 106187. (https://doi.org/10.1016/j.jastp.2024.106187)
65. 'Rebuilding the Ozone Layer: How the World Came Together for the Ultimate Repair Job', UN Environment Programme, 15 September 2021. (https://www.unep.org/news-and-stories/story/rebuilding-ozone-layer-how-world-came-together-ultimate-repair-job)

66. Priday, R. 'A Volcano-Inspired Weapon to Fix Climate Change is a Terrible Idea', Wired, 8 August 2018. (https://www.wired.com/story/volcanoes-climate-change-study-sulphur-stratosphere/)
67. Temple, J. ''The US Government is Developing a Solar Geoengineering Research Plan', *MIT Technology Review*, 1 July 2022. (https://www.technologyreview.com/2022/07/01/1055324/the-us-government-is-developing-a-solar-geoengineering-research-plan/)
68. Flavelle, C. and Gelles, D. 'U.K. to Fund "Small-Scale" Outdoor Geoengineering Tests', *New York Times*, 13 September 2024. (https://www.nytimes.com/2024/09/13/climate/united-kingdom-geoengineering-research.html)
69. 'Global Sulphur Dioxide (SO_2) Emissions by World Region', Our World in Data, accessed 15 February 2025. (https://ourworldindata.org/grapher/so-emissions-by-world-region-in-million-tonnes)
70. 'Sulfur Dioxide Basics', US Environmental Protection Agency, 10 January 2025. (https://www.epa.gov/so2-pollution/sulfur-dioxide-basics)
71. Weisenstein, D. K. et al. 'An Interactive Stratospheric Aerosol Model Intercomparison of Solar Geoengineering by Stratospheric Injection of SO_2 or Accumulation-Mode Sulfuric Acid Aerosols', *Atmospheric Chemistry and Physics*, 22 (5) (4 March 2022), pp. 2955–73. (https://doi.org/10.5194/acp-22-2955-2022)
72. Song, A. 'From Pollution to Solution', Cremieux Recueil, 25 May 2024. (https://www.cremieux.xyz/p/from-pollution-to-solution)
73. 'FAQ', Make Sunsets, accessed 15 February 2025. (https://makesunsets.com/pages/faq)

CHAPTER 7: SUPERCRITICAL

1. Wismayer, H. 'Searching For Climate Salvation In Deep Hellfire', *Noema*, 23 January 2025. (https://www.noemamag.com/searching-for-climate-salvation-in-deep-hellfire)
2. Zhao, Y. et al. 'Deformation and Instability Failure of Borehole at High Temperature and High Pressure in Hot Dry Rock Exploitation', *Renewable Energy*, 77 (1 May 2015), pp. 159–65. (https://doi.org/10.1016/j.renene.2014.11.086)
3. Crawford, J. 'Why Has Nuclear Power Been a Flop?', The Roots of Progress, 16 April 2021. (https://blog.rootsofprogress.org/devanney-on-the-nuclear-flop)

4. Reinhardt, B. 'Making Energy Too Cheap to Meter', *Works in Progress*, 12 October 2022. (https://worksinprogress.co/issue/making-energy-too-cheap-to-meter/)
5. Estrada, M. 'Radiogenic Heat', Stanford University, 22 March 2015. (http://large.stanford.edu/courses/2014/ph241/estrada1/)
6. Sistiaga, A. et al, 'Microbial Biomarkers Reveal a Hydrothermally Active Landscape at Olduvai Gorge at the Dawn of the Acheulean, 1.7 Ma', *Proceedings of the National Academy of Sciences of the United States of America*, vol. 117 No. 40 (14 August 2020. (https://www.pnas.org/doi/10.1073/pnas.2004532117)
7. 'Electricity Data Browser', US Energy Information Administration, accessed 15 February 2025. (https://www.eia.gov/electricity/data/browser/)
8. 'Installed Geothermal Energy Capacity, 2023', Our World in Data, 1 November 2024. (https://ourworldindata.org/grapher/installed-geothermal-capacity)
9. Lackovic, M. et al. 'Accelerating Decarbonization with Next-Generation Geothermal', BCG Global, 9 December 2022. (https://www.bcg.com/publications/2022/the-advantages-of-geothermal-energy)
10. Romanov, D. and Leiss, B. 'Geothermal Energy at Different Depths for District Heating and Cooling of Existing and Future Building Stock', *Renewable and Sustainable Energy Reviews* 167 (1 October 2022), 112727. (https://doi.org/10.1016/j.rser.2022.112727)
11. 'Geothermal', International Renewable Energy Agency, accessed 15 February 2025. (https://www.irena.org/Energy-Transition/Technology/Geothermal-energy)
12. 'Cost of Electricity in Texas', EnergySage, accessed 15 February 2025. (https://www.energysage.com/local-data/electricity-cost/tx/)
13. Thumm, M. 'Present State-of-the Art of the Worldwide Fusion Gyrotron Development', in GeMiC 2014: *German Microwave Conference*, 2014. (https://ieeexplore.ieee.org/document/6775118)
14. Potter, B. 'The Technological Innovations That Produced the Shale Revolution', IFP, 30 October 2023. (https://ifp.org/the-technological-innovations-that-produced-the-shale-revolution/)
15. Penrod, E. 'NV Energy Seeks New Tariff to Supply Google with 24/7 Power from Fervo Geothermal Plant', Utility Dive, 21 June 2024. (https://www.utilitydive.com/news/google-fervo-nv-energy-nevada-puc-clean-energy-tariff/719472/)

16. Hook, L. 'UK's First Deep Geothermal Energy Project for 37 Years Switched on', *Financial Times*, 18 June 2023. (https://www.ft.com/content/774c05ac-373d-45dc-bcaa-8377ddacb58f)
17. 'About the Eden Geothermal Project', Eden Geothermal, accessed 15 February 2025. (https://www.edengeothermal.com/the-project/phase-1/)
18. Vergnault, O. 'Earthquake Near Eden Project Was One of 300 Man-Made "Seismic Events" Since January', Cornwall Live, 10 March 2022. (https://www.cornwalllive.com/news/cornwall-news/earthquake-near-eden-project-one-6781365)
19. Cobham, T. 'Map: How Rare Are UK Earthquakes? Country "Riddled" with Fault Lines', *Independent*, 29 June 2023. (https://www.independent.co.uk/news/uk/home-news/earthquake-map-uk-west-midlands-b2366323.html)
20. Turak, N. 'Iran Says It's Discovered What Could Be the World's Second-Largest Lithium Deposit', CNBC, 6 March 2023. (https://www.cnbc.com/2023/03/06/iran-says-its-discovered-worlds-second-largest-lithium-deposit.html)
21. Andersen, R. 'America Just Hit the Lithium Jackpot', *The Atlantic*, 14 September 2023. (https://www.theatlantic.com/technology/archive/2023/09/nevada-lithium-geopolitics/675325/)
22. Lateef, S. '"As Geopolitically Important as Oil": Lithium Mining Could Transform India – but is Trouble Looming?', *Telegraph*, 15 June 2023. (https://www.telegraph.co.uk/global-health/climate-and-people/lithium-mining-india-precious-metal-batteries-carbon/)

CHAPTER 8: THE MISFIT PRODIGY

1. 'Future of Humanity Institute', futureofhumanityinstitute.org, 17 April 2024. (https:/www.futureofhumanityinstitute.org/)
2. Khatchadourian, R. 'The Doomsday Invention', *New Yorker*, 16 November 2015. (https://www.newyorker.com/magazine/2015/11/23/doomsday-invention-artificial-intelligence-nick-bostrom)
3. Bostrom, N. 'Nick Bostrom – Curriculum Vitae', accessed 15 February 2025. (https://nickbostrom.com/cv.pdf)
4. Ibid.
5. Bostrom, N. 'Existential Risks: Analyzing Human Extinction Scenarios and Related Hazards', *Journal of Evolution and Technology*, 9 (2002). (https://nickbostrom.com/existential/risks.pdf)

6. Bostrom, N. 'Are You Living in a Computer Simulation?', *Philosophical Quarterly*, 53 (211) (2003), pp. 243–255. (https://simulation-argument.com/simulation.pdf)
7. Ibid.
8. Ibid.
9. Williamson, M. 'Dr James Martin: Technology Guru and Philanthropist Who Predicted the Internet', *Independent*, 28 June 2013. (https://www.independent.co.uk/news/obituaries/dr-james-martin-technology-guru-and-philanthropist-who-predicted-the-rise-of-the-internet-8679451.html)
10. 'Dr James Martin – Our Founder', Oxford Martin School, accessed 15 February 2025. (https://www.oxfordmartin.ox.ac.uk/about/founder)
11. Edmonds, D. *Parfit: A Philosopher and His Mission to Save Morality*, Princeton, Princeton University Press, 2023.
12. Bostrom, N. and Sandberg, A. 'Whole Brain Emulation: A Roadmap', Future of Humanity Institute, Oxford University, 2008. (https://www.fhi.ox.ac.uk/brain-emulation-roadmap-report.pdf)
13. Bostrom, N. and Yudkowsky, E. 'The Ethics of Artificial Intelligence', in Ramsey, W. and Frankish, K. (eds). *Cambridge Handbook of Artificial Intelligence*, Cambridge University Press, 2011. (https://nickbostrom.com/ethics/artificial-intelligence.pdf)
14. Macaskill, W. 'Getting Inspired by Cost-Effective Giving', The Life You Can Save, 20 May 2013. (https://www.thelifeyoucansave.org/supporters-stories/getting-inspired-by-cost-effective-giving/)
15. Regis, E. 'The Incredible Shrinking Man', *Wired*, 1 October 2004. (https://www.wired.com/2004/10/drexler/)
16. Sandberg, A. 'Future of Humanity Institute 2005–2024: Final Report', University of Oxford, accessed 16 February 2025. (https://static1.squarespace.com/static/660e95991cf0293c2463bcc8/t/661a3fc3cecceb2b8ffce80d/1712996303164/FHI+Final+Report.pdf)
17. Sandberg, A. 'Andart II – Part of Anders' Exoself', aleph.se, 30 October 2021. (http://aleph.se/andart2/)
18. Bostrom, N. 'The Unfinished Fable of the Sparrows', OUP blog, 29 August 2014. (https://blog.oup.com/2014/08/unfinished-fable-sparrows-superintelligence/)
19. 'Elon Musk Funds Oxford and Cambridge University Research on Safe and Beneficial Artificial Intelligence', The Future of Humanity Institute, 1 July 2015. (https://www.fhi.ox.ac.uk/elon-musk-funds-oxford-and-cambridge-university-research-on-safe-and-beneficial-artificial-intelligence/)

20. Altman, S. 'Machine Intelligence, Part 1', blog.samaltman.com, 25 February 2015. (https://blog.samaltman.com/machine-intelligence-part-1)
21. 'Nick Bostrom Sets out Threats from Future Technologies at UN Meeting', Oxford Martin School, 12 October 2015. (https://www.oxfordmartin.ox.ac.uk/news/201510-bostrom-crbn-un)
22. Lin, S, et al. 'TruthfulQA: Measuring How Models Mimic Human Falsehoods', arXiv, 8 May 2022. (https://doi.org/10.48550/arXiv.2109.07958)
23. 'Home', AI Impacts, 19 December 2014. (https://aiimpacts.org/)
24. 'Future of Humanity Institute 2005–2024: Final Report', Effective Altruism Forum, 17 April 2024. (https://forum.effectivealtruism.org/posts/uK27pds7J36asqJPt/future-of-humanity-institute-2005-2024-final-report)
25. Lewis-Kraus, G. 'The Reluctant Prophet of Effective Altruism', *New Yorker*, 8 August 2022. (https://www.newyorker.com/magazine/2022/08/15/the-reluctant-prophet-of-effective-altruism)
26. MacAskill, W. 'Longtermism', Effective Altruism Forum, 25 July 2019. (https://forum.effectivealtruism.org/posts/qZyshHCNkjs3TvSem/longtermism)
27. Sandberg, A. 'Future of Humanity Institute 2005–024: Final Report', accessed 16 February 2025
28. Moorhouse, F. 'Goodbye to FHI', accessed 16 February 2025. (https://finmoorhouse.com/writing/fhi/)
29. Bostrom, N. 'Poetry', accessed 16 February 2025. (https://nickbostrom.com/poetry/poetry)
30. 'FHI at Oxford (Feat. Nick Bostrom)', YouTube, 2024. (https://www.youtube.com/watch?v=YLHS43Wo1As)
31. Robins-Early, N. 'Oxford Shuts down Institute Run by Elon Musk-Backed Philosopher', *Guardian*, 19 April 2024. (https://www.theguardian.com/technology/2024/apr/19/oxford-future-of-humanity-institute-closes)
32. 'Express Interest in an "FHI of the West"', LessWrong, 18 April 2024. (https://www.lesswrong.com/posts/ydheLNeWzgbco2FTb/express-interest-in-an-fhi-of-the-west)
33. Overcoming Bias, 5 January 2007. (https://web.archive.org/web/20070105153523/https:/www.overcomingbias.com/)
34. Hale, T. et al. 'Toward a Declaration on Future Generations', Blavatnik School of Government, University of Oxford, 12 January 2023. (http://www.bsg.ox.ac.uk/research/publications/toward-declaration-future-generations)

35. Bostrom, N. *Deep Utopia: Life and Meaning in a Solved World*, IdeaPress Publishing, 2024.

CHAPTER 9: DISEASE X

1. Brenner, D. 'A New Weapon in the Fight Against Superbugs', TED, YouTube, 2018. (https://www.youtube.com/watch?v=YATYsgi3e5A)
2. Wang, H. et al. 'Estimating Excess Mortality Due to the COVID-19 Pandemic: A Systematic Analysis of COVID-19-Related Mortality, 2020–21', *The Lancet*, 399 (10334) (16 April 2022), pp. 1513–36. (https://doi.org/10.1016/S0140-6736(21)02796-3)
3. Buonanno, M. et al. 'Far-UVC Light (222 Nm) Efficiently and Safely Inactivates Airborne Human Coronaviruses', *Scientific Reports*, 10 (1) (24 June 2020), 10285. (https://doi.org/10.1038/s41598-020-67211-2)
4. '222nm Far UVC Light Sterilizers Ultraviolet C Sanitizer Disinfection Lamp Suit for Schools Buses Airports Hospitals Bars Offices Homes Hotels(20W) …', Amazon, accessed 16 February 2025. (https://www.amazon.co.uk/Sterilizers-Ultraviolet-Sanitizer-Disinfection-Hosptails/dp/B08LBLSL2X?nsdOptOutParam=true&ufe=app_do:amzn1.fos.d7e5a2de-8759-4da3-993c-d11b6e3d217f)
5. Berche, P, 'Life and Death of Smallpox', *Presse Medicale*, 51 (3) (September 2022), 104117. (https://doi.org/10.1016/j.lpm.2022.104117)
6. Tucker, J. B. 'Biological Weapons in the Former Soviet Union: An Interview with Dr. Kenneth Alibek', *The Nonproliferation Review*, September 1999. (https://doi.org/10.1080/10736709908436760)
7. 'Bioterrorism Beginnings: The Rajneesh Cult, Oregon, 1985', OUP blog, 5 October 2009. (https://blog.oup.com/2009/10/bioterrorism-beginnings/)
8. Hughes, J. M. and Gerberding, J. L. 'Anthrax Bioterrorism: Lessons Learned and Future Directions', *Emerging Infectious Diseases*, 8 (10) (October 2002), pp. 1013–14. (https://doi.org/10.3201/eid0810.020466)
9. Zietz, B. P. and Dunkelberg, H. 'The History of the Plague and the Research on the Causative Agent Yersinia Pestis', *International Journal of Hygiene and Environmental Health*, 207 (2) (1 January 2004), pp. 165–78. (https://doi.org/10.1078/1438-4639-00259)

10. Nowak, R. 'Killer Mousepox Virus Raises Bioterror Fears', *New Scientist*, 10 January 2001. (https://www.newscientist.com/article/dn311-killer-mousepox-virus-raises-bioterror-fears/)
11. Koblentz, G. D. 'The De Novo Synthesis of Horsepox Virus: Implications for Biosecurity and Recommendations for Preventing the Reemergence of Smallpox', *Health Security*, 15 (6) (2017), pp. 620–28. (https://doi.org/10.1089/hs.2017.0061)
12. 'Products', kilobaser.com, accessed 16 February 2025. (https://kilobaser.com/webshop)
13. Carter, S. R. et al. 'Benchtop DNA Synthesis Devices: Capabilities, Biosecurity Implications, and Governance', Nuclear Threat Initiative, May 2023. (https://www.nti.org/wp-content/uploads/2023/05/NTIBIO_Benchtop-DNA-Report_FINAL.pdf)
14. Meselson, M. et al. 'The Sverdlovsk Anthrax Outbreak of 1979', *Science*, 266 (5188) (18 November 1994), pp. 1202–8. (https://doi.org/10.1126/science.7973702)
15. Board on Life Sciences et al. 'Gain-of-Function Research: Background and Alternatives', in *Potential Risks and Benefits of Gain-of-Function Research: Summary of a Workshop*, National Academies Press, 2015. (https://www.ncbi.nlm.nih.gov/books/NBK285579/)
16. 'Global Biolabs', globalbiolabs.org, accessed 16 February 2025. (https://www.globalbiolabs.org)
17. Herfst, S. et al. 'Airborne Transmission of Influenza A/H5N1 Virus between Ferrets', *Science*, 336 (6088) (22 June 2012), pp. 1534–41. (https://doi.org/10.1126/science.1213362)
18. Ibid.
19. Young A. and Vanden Brook, T. 'Pentagon: Poor Testing Led to Army Shipping Live Anthrax', *USA Today*, 23 July 2015. (https://www.usatoday.com/story/news/nation/2015/07/23/army-anthrax-shipments-pentagon-army/30154545/)
20. Penzenstadler, N. 'State Incidents Highlight Bioterror Lab Concerns', *Post-Crescent*, 28 May 2015. (https://www.postcrescent.com/story/news/investigations/2015/05/28/state-incidents-highlight-bioterror-lab-concerns/28089943/)
21. Director of Laboratory Science and Safety, FDA, 'FDA Review of the 2014 Discovery of Vials Labeled "Variola" and Other Vials Discovered in an FDA-Occupied Building on the NIH Campus', 13 December 2016. (https://www.fda.gov/media/101811/download)

22. Graves Fitzsimmons, E. 'Researcher Had Bacteria for Plague at His Death', *New York Times*, 21 September 2009. (https://www.nytimes.com/2009/09/22/us/22chicago.html)
23. 'What is the Legacy of the 2001 Foot-and-Mouth Outbreak?', BBC News, 18 February 2011. (https://www.bbc.com/news/uk-england-12472230)
24. Coghlan, A. 'Faulty Pipe Blamed for UK Foot and Mouth Outbreak', *New Scientist*, 7 September 2007. (https://www.newscientist.com/article/dn12615-faulty-pipe-blamed-for-uk-foot-and-mouth-outbreak/)
25. Barnes, J. E. 'Lab Leak Most Likely Caused Pandemic, Energy Dept. Says', *New York Times*, 26 February 2023. (https://www.nytimes.com/2023/02/26/us/politics/china-lab-leak-coronavirus-pandemic.html)
26. 'IMF Sees Cost of COVID Pandemic Rising beyond $12.5 Trillion Estimate', Reuters, 20 January 2022. (https://www.reuters.com/business/imf-sees-cost-covid-pandemic-rising-beyond-125-trillion-estimate-2022-01-20/)
27. Tong, C. et al. 'Chlorine Disinfectants Promote Microbial Resistance in Pseudomonas sp.', *Environmental Research*, 199 (August 2021), 111296. (https://doi.org/10.1016/j.envres.2021.111296)
28. Matulonga, B. et al. 'Women Using Bleach for Home Cleaning are at Increased Risk of Non-Allergic Asthma', *Respiratory Medicine*, 117 (1 August 2016), pp. 264–71. (https://doi.org/10.1016/j.rmed.2016.06.019)
29. Arendsen, L. P. et al. 'The Use of Copper as an Antimicrobial Agent in Health Care, Including Obstetrics and Gynecology', *Clinical Microbiology Reviews*, 32 (4) (14 August 2019). (https://pmc.ncbi.nlm.nih.gov/articles/PMC6730497/)
30. Schmidt, M. G. et al. 'Self-Disinfecting Copper Beds Sustain Terminal Cleaning and Disinfection Effects throughout Patient Care', *Applied and Environmental Microbiology*, 86 (1) (13 December 2019), e01886-19. (https://doi.org/10.1128/AEM.01886-19)
31. Salgado, C. D. et al. 'Copper Surfaces Reduce the Rate of Healthcare-Acquired Infections in the Intensive Care Unit', *Infection Control & Hospital Epidemiology*, 34 (5) (May 2013), pp. 479–86. (https://doi.org/10.1086/670207)
32. Craft, S. 'How Many Schools Are in the US? (Statistics 2025)', ThinkImpact, 8 November 2024. (https://www.thinkimpact.com/number-of-schools-in-the-us/)

33. Qi, S. et al. 'Porous Cellulose Thin Films as Sustainable and Effective Antimicrobial Surface Coatings', *ACS Applied Materials & Interfaces*, 15 (17) (2023), pp. 20638–48. (https://doi.org/10.1021/acsami.2c23251)
34. 'A Leader in Antimicrobial Innovation: Permanently Protecting against the Spread of Infectious Microbes', LuminOre CopperTouch, accessed 16 February 2025. (https://www.luminore-copper.com)
35. 'Cellulose Coating Neutralises COVID Virus in Minutes', *The Engineer*, 17 May 2023. (https://www.theengineer.co.uk/content/news/cellulose-coating-neutralises-covid-virus-in-minutes/)
36. 'The Cost of Sequencing a Human Genome', National Human Genome Research Institute, 1 November 2021. (https://www.genome.gov/about-genomics/fact-sheets/Sequencing-Human-Genome-cost)
37. Department of Health and Social Care et al. 'UK to Create World-First "Early Warning System" for Pandemics', GOV.UK, 5 November 2024. (https://www.gov.uk/government/news/uk-to-create-world-first-early-warning-system-for-pandemics)
38. Klapsa, D. et al. 'Sustained Detection of Type 2 Poliovirus in London Sewage between February and July, 2022, by Enhanced Environmental Surveillance', *The Lancet*, 400 (10362) (29 October 2022), pp. 1531–38. (https://doi.org/10.1016/S0140-6736(22)01804-9)
39. Mohring, J. et al. 'Estimating the COVID-19 Prevalence from Wastewater', *Scientific Reports*, 14 (1) (22 June 2024), 14384. (https://doi.org/10.1038/s41598-024-64864-1)
40. Ziady, H. 'Biotech Company Moderna Says Its Coronavirus Vaccine is Ready for First Tests', CNN, 26 February 2020. (https://www.cnn.com/2020/02/25/business/moderna-coronavirus-vaccine/index.html)
41. Garde, D. and Saltzman, J. 'The Story of mRNA: How a Once-Dismissed Idea Became a Leading Technology in the Covid Vaccine Race', STAT, 10 November 2020. (https://www.statnews.com/2020/11/10/the-story-of-mrna-how-a-once-dismissed-idea-became-a-leading-technology-in-the-covid-vaccine-race/)
42. Binday, B. '"Not of Faculty Quality": How Penn Mistreated Nobel Prize-Winning Researcher Katalin Karikó', *Daily Pennsylvanian*, 26 October 2023. (https://www.thedp.com/article/2023/10/penn-katalin-kariko-university-relationship-mistreatment)

43. 'Moderna Vaccine Slightly More Effective than Pfizer Vaccine in Preventing COVID-19 Infection, Hospitalization, and Death', Harvard T.H. Chan School of Public Health, 1 December 2021. (https://hsph.harvard.edu/news/moderna-vaccine-slightly-more-effective-than-pfizer-vaccine/)
44. Wang, B. 'Day Zero Antivirals for Future Pandemics', Asimov Press, 26 August 2024. (https://press.asimov.com/articles/antivirals)
45. 'BioNTech and CEPI Expand Partnership to Strengthen Africa's mRNA Vaccine Ecosystem', CEPI, 29 May 2024. (https://cepi.net//biontech-and-cepi-expand-partnership-strengthen-africas-mrna-vaccine-ecosystem)

CHAPTER 10: THE WEST SIBERIAN HAT FACTORY – AND WORSE

1. Lentzos, F. 'What Happened after an Explosion at a Russian Disease Research Lab Called VECTOR?, *Bulletin of the Atomic Scientists*, 27 November 2019. (https://thebulletin.org/2019/11/what-happened-after-an-explosion-at-a-russian-disease-research-lab-called-vector/)
2. Hoffman, p. 103.
3. Ibid. pp. 105–6.
4. 'United States Biological Overview', The Nuclear Threat Initiative, 4 June 2015. (https://www.nti.org/analysis/articles/united-states-biological/)
5. Pontin, M. W. 'The Knowledge', *MIT Technology Review*, 1 March 2006. (https://www.technologyreview.com/2006/03/01/229580/the-knowledge/)
6. Hoffman, p. 110.
7. Ibid., p. 473.
8. 'Report of the World Health Organisation (WHO) Biosafety Inspection Team of the Variola Virus Maximum Containment Laboratories to the State Research Centre of Virology and Biotechnology ("SRC VB VECTOR"), Federal Service for Surveillance on Consumer Rights Protection and Human Well-Being, Novosibirsk', World Health Organisation, 2009. (https://cdn.who.int/media/docs/default-source/documents/health-topics/smallpox-vector-inspection-20094e8f2622-058f-4931-9f9c-f0add1240448.pdf?sfvrsn=367f02f3_1)
9. Field, M. 'An Explosion Rocked a Russian Research Facility Known for Housing the Smallpox Virus', *Bulletin of the Atomic Scientists*,

16 September 2019. (https://thebulletin.org/2019/09/an-explosion-rocked-a-russian-research-facility-known-for-housing-the-smallpox-virus/)

10. US Department of State, 'Adherence to and Compliance with Arms Control, Nonproliferation, and Disarmament Agreements and Commitments', April 2023. (https://www.state.gov/wp-content/uploads/2023/04/13APR23-FINAL-2023-Treaty-Compliance-Report-UNCLASSIFIED-UNSOURCED.pdf)
11. Marani, M. et al. 'Intensity and Frequency of Extreme Novel Epidemics', *Proceedings of the National Academy of Sciences* 118 (35) (31 August 2021), e2105482118. (https://doi.org/10.1073/pnas.2105482118)
12. Miller, J. 'From Bioweapons Lab, a Formula for Peace', *Los Angeles Times*, 13 August 2006. (https://www.latimes.com/archives/la-xpm-2006-aug-13-oe-jmiller13-story.html)
13. 'Containment Facilities as Explained in the Biosafety in Microbiological and Biomedical Laboratories of the US National Institutes of Health's Centers for Disease Control and Prevention', accessed 16 February 2025. (https://static1.squarespace.com/static/62fa334a3a6fe8320f5dcf7e/t/649447accc4f547bb0783481/1687439276489/Containment+Facilities+explained+in+the+BMBL+6th+ed.pdf)
14. Tétrault-Farber, G. and Farge, E. '"High Bio-Hazard Risk" in Sudan after Laboratory Seized, WHO Says', Reuters, 25 April 2023. (https://www.reuters.com/world/africa/un-humanitarian-office-reduces-sudan-operations-due-fighting-2023-04-25/)
15. 'Global BioLabs Report 2023', King's College London and George Mason University, accessed 16 February 2025. (https://static1.squarespace.com/static/62fa334a3a6fe8320f5dcf7e/t/6412d3120ee69a4f4efbec1f/1678955285754/KCL0680_BioLabs+Report_Digital.pdf)
16. 'ISO 35001:2019', ISO, 2019. (https://www.iso.org/standard/71293.html)
17. 'Czech Uranium Removed', Arms Control Association, accessed 16 February 2025. (https://www.armscontrol.org/act/2005-11/czech-uranium-removed)
18. Kane, A. et al. 'Joint Assessment Mechanism to Determine Pandemic Origins', Nuclear Threat Initiative, accessed 16 February 2025. (https://www.nti.org/about/programs-projects/project/joint-assessment-mechanism-to-determine-pandemic-origins/)

19. Carter, S. R. et al. 'The Convergence of Artificial Intelligence and the Life Sciences', Nuclear Threat Initiative, October 2023. (https://www.nti.org/wp-content/uploads/2023/10/NTIBIO_AI_FINAL.pdf)
20. Adamala, K. P. et al. 'Confronting Risks of Mirror Life', *Science*, 386 (6728) (20 December 2024), pp. 1351–53. (https://doi.org/10.1126/science.ads9158)
21. 'How AI Takeover Might Happen in 2 Years', LessWrong, 7 February 2025. (https://www.lesswrong.com/posts/KFJ2LFogYqzfGB3uX/how-ai-takeover-might-happen-in-2-years)

CHAPTER 11: THE RACE FOR ALIGNMENT

1. Yudkowsky, E. 'The AI Alignment Problem: Why It's Hard, and Where to Start', Machine Intelligence Research Institute, 5 May 2016. (https://intelligence.org/stanford-talk/)
2. 'OpenAI Email Archives (from Musk v. Altman and OpenAI Blog)', LessWrong, 16 November 2024. (https://www.lesswrong.com/posts/5jjk4CDnj9tA7ugxr/openai-email-archives-from-musk-v-altman-and-openai-blog)
3. Metz, C. 'Teaching A.I. Systems to Behave Themselves', *New York Times*, 13 August 2017. (https://www.nytimes.com/2017/08/13/technology/artificial-intelligence-safety-training.html)
4. Patel, D. 'Paul Christiano – Preventing AI Takeover', Dwarkesh podcast, 31 October 2023. (https://www.dwarkeshpatel.com/p/paul-christiano)
5. Bloom, R. and Habryka, O. 'Welcome & FAQ!', AI Alignment Forum, 24 August 2021. (https://www.alignmentforum.org/about)
6. Henshall, W. and Perrigo, B. 'TIME100 AI 2023: Dario and Daniela Amodei', *TIME*, 7 September 2023. (https://time.com/collection/time100-ai/6309047/daniela-and-dario-amodei/)
7. Bai, Y. et al. 'Constitutional AI: Harmlessness from AI Feedback', arXiv, 15 December 2022. (https://doi.org/10.48550/arXiv.2212.08073)
8. 'Introducing Claude', Anthropic, 14 March 2023. (https://www.anthropic.com/news/introducing-claude)
9. Field, M. 'From Black Nazis to Female Popes and American Indian Vikings: How AI Went "Woke"', *Telegraph*, 23 February 2024. (https://www.telegraph.co.uk/news/2024/02/23/google-gemini-ai-images-wrong-woke/)

10. Yudkowsky, E. 'Will Superintelligent AI End the World?', TED, April 2023. (https://www.ted.com/talks/eliezer_yudkowsky_will_superintelligent_ai_end_the_world/transcript)
11. Sharkey, L. 'Increasing Access to Pain Relief in Developing Countries – An EA Perspective', Effective Altruism Forum, 31 January 2017. (https://forum.effectivealtruism.org/posts/nsCFsRjfW6b72gJuB/increasing-access-to-pain-relief-in-developing-countries-an)
12. Greenblatt, R. et al. 'AI Control: Improving Safety Despite Intentional Subversion', arXiv, 23 July 2024. (https://doi.org/10.48550/arXiv.2312.06942)
13. Goldman, S. 'NIST Staffers Revolt against Expected Appointment of "Effective Altruist" AI Researcher to US AI Safety Institute', VentureBeat, 7 March 2024. (https://venturebeat.com/ai/nist-staffers-revolt-against-potential-appointment-of-effective-altruist-ai-researcher-to-us-ai-safety-institute/)
14. Gebru, T. 'Effective Altruism Is Pushing a Dangerous Brand of "AI Safety"', *Wired*, 20 November 2022. (https://www.wired.com/story/effective-altruism-artificial-intelligence-sam-bankman-fried/)
15. Ibid.
16. Hutson, M. 'AI Learns the Art of Diplomacy', science.org, 22 November 2022. (https://www.science.org/content/article/ai-learns-art-diplomacy-game)
17. Jackson, S. 'The CEO of the Company Behind AI Chatbot ChatGPT Says the Worst-Case Scenario for Artificial Intelligence is "Lights out for All of Us"', *Business Insider*, 4 July 2023. (https://www.businessinsider.com/chatgpt-openai-ceo-worst-case-ai-lights-out-for-all-2023-1)
18. Roose, K. 'Silicon Valley Confronts a Grim New A.I. Metric', *New York Times*, 6 December 2023. (https://www.nytimes.com/2023/12/06/business/dealbook/silicon-valley-artificial-intelligence.html)
19. Ibid.

CHAPTER 12: POKING THE SHOGGOTH

1. ''Voting Intention Polls and Trends', Ipsos, 4 July 2024. (https://www.ipsos.com/en-uk/uk-opinion-polls/voting-intention-polls-and-trends)
2. Science, Innovation and Technology Committee, 'Governance of Artificial Intelligence (AI)', UK Parliament, 28 May 2024. (https://

publications.parliament.uk/pa/cm5804/cmselect/cmsctech/38/report.html)

3. Roose, K. 'A Conversation With Bing's Chatbot Left Me Deeply Unsettled', *New York Times*, 16 February 2023. (https://www.nytimes.com/2023/02/16/technology/bing-chatbot-microsoft-chatgpt.html)
4. Hurler, K. 'Chat-GPT Pretended to Be Blind and Tricked a Human Into Solving a CAPTCHA', Gizmodo, 15 March 2023. (https://gizmodo.com/gpt4-open-ai-chatbot-task-rabbit-chatgpt-1850227471)
5. Barnes, B. 'More Information About the Dangerous Capability Evaluations We Did with GPT-4 and Claude', LessWrong, 19 March 2023. (https://www.lesswrong.com/posts/4Gt42jX7RiaNaxCwP/more-information-about-the-dangerous-capability-evaluations)
6. Henshall, W. 'Nobody Knows How to Safety-Test AI', *TIME*, 21 March 2024. (https://time.com/6958868/artificial-intelligence-safety-evaluations-risks/)
7. O'Brien, M. and Parvini, S. 'How Harris and Trump Differ on Artificial Intelligence Policy', Associated Press, 30 July 2024. (https://apnews.com/article/ai-artificial-intelligence-kamala-harris-trump-vance-1c94ba2ee6a1410e61d980faa51b19e4)

CHAPTER 13: **WHISTLEBLOWERS**

1. Brundage, M. 'Why I'm Leaving OpenAI and What I'm Doing Next', Miles's Substack, 23 October 2024. (https://milesbrundage.substack.com/p/why-im-leaving-openai-and-what-im)
2. Miles Brundage [@Miles_Brundage], 'I Just Sent This Message to My Colleagues, and Elaborate on My Decision and next Steps in a Blog Post (See next Tweet)', Twitter post, 23 October 2024. (https://x.com/Miles_Brundage/status/1849138802864087234)
3. Kokotajlo, D. 'What 2026 Looks Like', LessWrong, 6 August 2021. (https://www.lesswrong.com/posts/6Xgy6CAf2jqHhynHL/what-2026-looks-like)
4. Henshall, W. 'When Might AI Outsmart Us? It Depends Who You Ask', *TIME*, 19 January 2024. (https://time.com/6556168/when-ai-outsmart-humans/)
5. Karger, E. et al. 'Forecasting Existential Risks: Evidence from a Long-Run Forecasting Tournament', *Forecasting Research Institute*, 8 August 2023. (https://static1.squarespace.com/

static/635693acf15a3e2a14a56a4a/t/64f0a7838ccbf43b6b5ee40c/1693493128111/XPT.pdf#page=697)

6. Cuthbertson, A. 'ChatGPT Boss Says He's Achieved Human-Level AI, Then Says He's "Just Memeing"', *Independent*, 27 September 2023. (https://www.independent.co.uk/tech/chatgpt-ai-agi-sam-altman-openai-b2419449.html)
7. Lawler, R. 'Former OpenAI Board Member Explains Why They Fired Sam Altman', The Verge, 29 May 2024. (https://www.theverge.com/2024/5/28/24166713/openai-helen-toner-explains-why-sam-altman-was-fired)
8. Michaël Trazzi [@MichaelTrazzi], '> Be Daniel Kokotajlo > Discover That AGI Is Imminent > Post Short Timeline Scenarios > Entire World Is Shocked > Go to OpenAI to Check Timelines > Find out You Were Correct > Job Done, Leave OpenAI > Give up 85% of Net Worth to Be Able to Criticize OpenAI > You're Actually The ...', Twitter, 23 May 2024. (https://x.com/MichaelTrazzi/status/1793669715971510630)
9. Hu, K. and Cai, K. 'Exclusive: OpenAI to Remove Non-Profit Control and Give Sam Altman Equity', Reuters, 26 September 2024. (https://www.reuters.com/technology/artificial-intelligence/openai-remove-non-profit-control-give-sam-altman-equity-sources-say-2024-09-25/)
10. Yudkowsky, E. 'Pausing AI Developments Isn't Enough. We Need to Shut It All Down', *TIME*, 29 March 2023. (https://time.com/6266923/ai-eliezer-yudkowsky-open-letter-not-enough/)
11. 'OpenAI and Elon Musk', OpenAI, 13 March 2024. (https://openai.com/index/openai-elon-musk/)

CHAPTER 14: THE METHUSELARITY

1. Keynes, J. M. *A Tract on Monetary Reform*, London, Macmillan, 1923.
2. Bostrom, N. and Ćirković, M. M. (eds.) *Global Catastrophic Risks*, Oxford, Oxford University Press, 2008.
3. 'Voronoff Predicts Active Life of 125 Years, Three Months "Old Age" and Then Death', *New York Times*, 17 March 1927. (https://www.nytimes.com/1927/03/17/archives/voronoff-predicts-active-life-of-125-years-three-months-old-age-and.html)
4. Friend, T. 'Silicon Valley's Quest to Live Forever', *New Yorker*, 27 March 2017. (https://www.newyorker.com/magazine/2017/04/03/silicon-valleys-quest-to-live-forever)

5. Smith, C. L. 'Aubrey de Grey: We Don't Have to Get Sick as We Get Older', *Guardian*, 1 August 2010. (https://www.theguardian.com/technology/2010/aug/01/aubrey-de-grey-ageing-research)
6. 'Genescient's Methuselah Flies', Genescient, accessed 16 February 2025. (https://www.genescient.com/research/methuselah-flies/)
7. Jensen, M. B. and LePort, F. 'Fruit Flies, Burning Man, and Creating Time: The Gordian Founding Story', Gordian Biotechnology, 18 April 2024. (https://www.gordian.bio/blog/gordian-founding-story/)
8. Ibid.

CONCLUSION: EXISTENTIAL HOPE

1. Cotton-Barratt, O. and Ord, T. 'Existential Risk and Existential Hope: Definitions', Future of Humanity Institute – Technical Report #2015-1, accessed 16 February 2025, (https://www.fhi.ox.ac.uk/Existential-risk-and-existential-hope.pdf)
2. Tolkien, J. R. R. *On Fairy-Stories*, Oxford, Oxford University Press, 1947.
3. Fisher, R. 'Eucatastrophe: Tolkien's Word for the "Anti-Doomsday"', 13 October 2022. (https://www.bbc.com/future/article/20221005-eucatastrophe-tolkiens-word-for-the-anti-doomsday)
4. Sandberg, A. 80,000 Hours '#165 – Anders Sandberg on war in space, whether civilisations age, and the best things possible in our universe'. (https://80000hours.org/podcast/episodes/anders-sandberg-best-things-possible-in-our-universe/)
5. Bregman, R. *Moral Ambition: Stop Wasting Your Talent and Start Making a Difference*, London, Bloomsbury, 2025.
6. Collins, M. *Carrying the Fire: An Astronaut's Journeys*, New York, Farrar, Straus & Giroux, 2009.

INDEX

INDEX